Cost-Benefit Analyses of Climate Change

The Broader Perspectives

F.L. Toth (ed.)

Springer Basel AG

Editor
Dr. Ferenc L. Toth
Potsdam Institute for Climate Impact Research
Dept. of Global Change & Social Systems
Telegrafenberg C4
P.O. Box 60 12 03
D-14412 Potsdam
Germany

Die Deutsche Bibliothek – CIP-Einheitsaufnahme
Cost benefit analyses of climate change : the broader perspectives /
F. L. Tóth (ed.) - Basel ; Boston ; Berlin : Birkhäuser, 1997
  ISBN 978-3-0348-9831-7

Library of Congress Cataloging-in-Publication Data
Cost-benefit analyses of climate change : the broader perspectives /
  F.L. Tóth. ed.
     p.   cm.
  Includes bibliographical references and index.
    ISBN 978-3-0348-9831-7   ISBN 978-3-0348-8928-5 (eBook)
    DOI 10.1007/978-3-0348-8928-5
     1. Climatic changes--Cost effectiveness--Congresses.   2. Global
  environmental change--International cooperation--Congresses.
  I. Tóth, F. L. (Ferenc L.), 1953-   .
  QC981.8.C5C76   1997
  338.4'336373874--dc21

© 1998 Springer Basel AG
Originally published by Birkhäuser Verlag AG in 1998
Softcover reprint of the hardcover 1st edition 1998
Cover design: Micha Lotrovsky, Therwil, Switzerland
Printed on acid-free paper produced from chlorine-free pulp. TCF ∞

ISBN 978-3-0348-9831-7

9 8 7 6 5 4 3 2 1

Table of contents

Part I: Models, concepts, and policy instruments

Beyond costs and benefits of climate change: A workshop overview
Ferenc L. Toth

First principles and the economic comparison of regulatory alternatives in global change
Gary W. Yohe

Dynamics of policy instruments and the willingness to participate in an international agreement
Jürgen E. Blank

Global warming and the insurance industry
Gerhard A. Berz

Part II. Stabilization targets, costs, and technologies

European stabilization targets: What do they bring, how much do they cost?
Eberhard Jochem

Climate protection and the economy of prevention
Peter Hennicke

The value of advanced energy technologies in stabilizing atmospheric CO_2
Jae Edmonds and Marshall Wise

Policy context: Follow-up of the Berlin climate conference
Michael Ernst

The implications of including sulfate aerosols on scenarios of admissible greenhouse gas emissions
Christian Issig, Hans-Jochen Luhmann and Paul Vradelis

The tolerable windows approach to climate control: Optimization, risks, and perspectives
Gerhard Petschel-Held and Hans-Joachim Schellnhuber

Foreword

The Potsdam Institute for Climate Impact Research (PIK) was founded in 1992 as a "Blue List" research institute, with the Federal Ministry for Education, Science, Research and Technology and the Ministry for Science, Research and Culture of the federal state of Brandenburg each providing half of the funding. PIK currently has a staff of 100 (1997), including about 75 scientists and guest scientists, as well as a number of students and temporary assistants. Further expansion is taking place at the institute site in the "Albert Einstein" Science Park in Potsdam.

The interdisciplinary nature of climate impact research, especially the interface between the natural scientific and socioeconomic dimensions of environmental research, is reflected at PIK in the close cooperation with partner institutes at national and international level. The flexible framework created for the institute enables new problems and issues to be taken up as they arise. As a center of scientific innovation, PIK also coordinates international activities in the fields of climate impact research and Earth System analysis. The institute houses project offices for the IGBP international research programs, for example. Simulations of Global Change are performed on PIK's supercomputer using models and data drawn from various disciplines. The parallel computer (an IBM-SP2) boasts 20 gigaflops of computing power, making it one of the most powerful research computers in Germany.

The Workshop on Cost-Benefit Analyses of Climate Change was jointly organized by PIK and the Wuppertal Institute for Energy, Climate, and Environment (WI). The workshop was held at PIK in Potsdam, Germany on 10–11 November 1995 and it was financially supported by the German Federal Ministry for Education and Research (BMBF). The workshop brought together leading scientists from Germany and selected researchers from around the world involved in studies on various aspects of cost-benefit assessments of global climate change. (See Appendix 1 for a list of participants.)

The main objective of the workshop was to explore major uncertain, unknown, or ignored items in integrated cost-benefit models of climate change in three major areas: modeling regional climate sensitivity, impacts, and adaptation; emissions reduction targets, associated costs, and possible instruments for implementation; and finally, the broader national and international policy contexts for using these models and their results. The goal was to generate creative discussions about current problems and possible future research directions.

The workshop was designed in a short introduction and 3 half-day sessions. Each session addressed 3-4 themes. Each thread covered one specific and largely unresolved issue in climate change. A thread was a focused discussion of the selected issue started by two invited presentations. The objective of each thread was to start out from where we are in understanding the issue at hand and propose/discuss possible directions for future research efforts. Accordingly, the presentations provided the necessary state-of-the-art review of the ways each issue is addressed in the current efforts of integrated cost-benefit modeling. (Appendix 2 includes the Workshop Program with the list of presentations.)

We asked leading experts in the field from Germany, the United States, and other countries to prepare a short paper based on their own presentations and the discussions at the workshop. Papers in this volume report about selected issues in this large field of scientific research and analyze the policy implications. Topics cover a broad range from the implications of including sulfate aerosols in target-based scenarios of climate change to impacts of global warming on the insurance industry.

On the policy side, contributors look at the costs and expected results of various greenhouse-gas stabilization strategies as well as the dynamics of policy instruments and the role of advanced energy technologies in those stabilization targets and strategies. In all these areas, papers address possible research directions that might help alleviate current shortcomings in global cost-benefit analyses and, more broadly, in integrated climate-economy assessment models.

In addition to the authors, many people contributed to producing this volume. Eva Hizsnyik coordinated the production process and served as technical editor. Comments from a large number of reviewers and thorough contribution by Thomas Bruckner, Hans-Martin Füssel, Carsten Helm and Marian Leimbach in the editing phase are gratefully acknowledged.

Hans-Joachim Schellnhuber, director
Ferenc L. Toth, project leader

Part I

Models, concepts, and policy instruments

Cost-Benefit Analyses of Climate Change: The Broader Perspective
F.L. Toth (ed.)
© 1998 Birkhäuser Verlag Basel/Switzerland

Beyond costs and benefits of climate change: A workshop overview

Ferenc L. Toth

PIK - Potsdam Institute for Climate Impact Research, Department of Global Change and Social Systems, P.O. Box 601203, D-14412 Potsdam, Germany

1. Introduction

Global climate change poses a major challenge for analysts and policy advisors alike. The complexities, uncertainties, and high stakes involved in the problem renders economic analysis particularly controversial. Many analysts maintain that analytical techniques developed over the past few decades and successfully adopted for managing a broad range of environmental issues can also be applied in the case of climate change, after making the changes necessary to accommodate special features of the climate problem. Others paint apocalyptic pictures of possible impacts of climate change on nature and society, and question the applicability of economic approaches altogether.

The relevance of cost-benefit analysis has been distinctively debated. Working Group III of the IPCC Second Assessment Report devoted an entire chapter to the subject. Unfortunately the statements are rather inconclusive. While the debate goes on, it is important to look beyond the narrow costs and benefits and look at the broader issues involved in the climate problem. Integrated assessments, and particularly integrated assessment models (IAMs), have emerged over recent years as the analytical framework to do just that.

Climate impact assessments belong to a distinct class of studies where an increasing scope of integration can be observed. The original single-sector approach (starting with scenarios of climate change, assessing their impact on individual biophysical systems and then their consequences for today's society and economy without considering adaptation options) has gradually become integrated along several criteria: they attempt to assess cross-sectoral impacts in the biophysical systems (soils, hydrology, vegetation), they now consider adaptation options in the socio-economic systems (local: farm, river basin, costal zone; and higher-level: regional and national level policies); they project climate-related vulnerabilities of future societies, and they increasingly rely on integrated biophysical and socio-economic assessments.

Perhaps the most prominent area of IAMs over the past few years is integrated climate-economy models in which assessments of impacts and damages from climate change and estimates of

i

the economic costs of slowing or preventing climate change are integrated into a single analytical framework. This framework also includes a simple representation of the major biogeochemical processes involved in climate change.

This paper is intended to provide a concise history of the modeling efforts in economics and a few closely related disciplines to provide the context and background to the papers collected in this volume. Section 2 looks at the evolution of the various components and their early linkage in the form of traditional cost-benefit assessments. The next two sections provide short descriptions of major model types in two major classes, simulation and optimization models. Section 5 then presents the layout for the book and highlights important features of the papers that follow.

2. An increasing scope of integration

In the initial period of several years, economic analyses of climate change proceeded in two virtually distinct areas. One group of researchers was trying to estimate the costs of the various greenhouse gas (GHG) abatement targets that had been proposed by various high-level international political meetings. A second group of analysts was trying to assess regional and local impacts, possibilities for adaptation, and a financial estimate of damage. These discrete research directions made results difficult to compare and proved to be of only limited use in the policy arena. There was a perceived need to move toward a consistent framework of analysis. This led to the first attempts to apply traditional cost-benefit analysis to the climate change issue.

Soon after, it was recommended that assessments of impact and damage from climate change and estimates of the economic costs of slowing or preventing climate change should be integrated into a single analytical framework, including a representation of major biogeochemical processes involved in climate change.

Over the past years, the number of models integrating the biophysical processes of climate and the most relevant processes of the economy has grown exponentially. These models analyse the full cycle of anthropogenic emissions of GHGs, their concentrations in the atmosphere, the resulting climate forcing, and finally the impact of the induced climatic change on the economy. For economists and social scientists, the great advantage of this approach is that it provides a comprehensive framework for assessing the possible economic losses due to climate change (damage function) and for estimating the costs to slow or delay climate change (cost function). By creating the same metric for cost and benefit assessments, integrated models can then be used to develop economically efficient climate policies.

2.1. Cost and benefit estimates

A closer look at the.early literature of economic analysis addressing the global warming issue shows major imbalances.

- First, the number of studies estimating the costs of GHG/CO_2 mitigation strategies is over-whelming compared to those assessing the benefits.
- Second, cost studies largely rely on rigorous analytical tools, in most cases a single, sophisti-cated and thoroughly tested computer model, as opposed to damage estimates that need to rely on a variety of fragmented, and in many cases self-contradictory impact assessments to derive an aggregated damage result. Impact assessment studies tend to focus on one specific crop in one region, inundation and property loss from sea level rise in another, and water resources in a third. Moreover, many studies analyse impacts under a 2*CO_2 equivalent climate and, at the same time, assume carbon fertilization effects of a 2*CO_2 concentration.
- Third, the evaluation of mitigating costs over time has proved to be possible by integrating the most important macroeconomic dynamics (GDP growth, productivity improvements, energy use) into the overall framework of analysis. This way, future costs of various abatement strategies are related to projected future GDP values and baseline emission trends. In contrast, impact assessments (and thus damage estimates) tend to focus on 2*CO_2 climate scenarios and superimpose them on present-day economic and technological conditions.
- Finally, and to a large extent explained by the previous points, the spread of results is much wider for damage assessments (in terms of potential GDP losses due to an assumed level of global warming) than for cost estimates (using the same terms to measure economic losses from a specified rate of reduction in CO_2 emissions). This seems to hold despite the apparent gap between results of top-down and bottom-up models applied in cost estimates.

No wonder that these imbalances invite considerable criticism when results are integrated into a benefit-cost framework. They also point toward the need for further improvements on both sides, and probably for profoundly new approaches on the damage side.

Beyond obvious differences in the multitude of factors to be considered and in the complexity of the analysis in damage vs cost estimates, one possible explanation for the above imbalances is that for the latter it was possible to extend earlier energy-economic models (both macroeconomic and engineering types) by incorporating additional constraints in the form of direct emission target levels or incentive-based instruments, like carbon taxes. Another motivation may have been that cost assessments and studies of different cost reduction schemes (permit trade, tax revenue recycling, etc.) will be important in their own right (and regardless of the damage results) if there is a policy decision to undertake emission mitigation strategies for other than purely economic reasons. It is difficult to tell whether it is due to the convenient availability of well-established models, or to the perceived need for these kinds of results even in the absence of full cost-benefit justification, but cost studies are certainly by far the most advanced area of greenhouse economics and modeling.

These imbalances also characterize integrated assessments. As will be seen in the next sections, GHG emissions in general and energy-CO_2 relationships in particular are represented in the integrated models in much greater detail than impacts and damages.

Benefits

The shape and relative position of the damage function, synthesizing measurable economic damages of climate change impacts, is determined by a broad range of geographical, socio-economic, and technological factors. The damage function is bound to change over time as those factors change even if the underlying climate change scenario remains the same.

The economic impacts of climate change are generally thought to be negative. Some negative impacts will be offset by positive effects in the same sector or economic activity, e.g. yield losses due to reduced maturation period partly offset by the atmospheric carbon fertilization effect, or part of the increased demand for space cooling in the summer will be offset by reduced costs of space heating in the winter if average temperature increases hold across the whole year evenly. Other negative impacts in one sector might be partly or fully compensated for by positive impacts in other sectors in the same national economy. Very few studies have dared to estimate the positive impacts of a warmer climate, and even fewer the economic benefits associated with them.

Considering the long-term trend of human activity becoming less vulnerable to climatic fluctuations and climate in general (Ausubel, 1991; Schelling, 1992), actual damages even in 30, but certainly in 50 to 100 years, will inevitably be lower than damages calculated by superimposing a future climate, whatever form it takes, on today's economy. The general pattern is global although if we assume a saturation pattern in increasing climate independence as a result of economic development, the autonomous rate of decrease in climatic vulnerability will be higher in developing countries, that is in regions thought to be the most vulnerable to climate change today.

The first serious and systematic effort to quantify economic damages from climate change (Nordhaus, 1991) seems to have become a benchmark or reference point for several other studies. Many authors criticize the Nordhaus estimates for its omissions and limited scope (see, for example, Cline, 1992, who also provides his own estimates which are somewhat higher than those of Nordhaus); others follow its basic principles and extend it to other world regions. One such extension is by Fankhauser (1993) to the global scale, which is interesting because some of his major world regions overlap or are reasonably close to the world regions used in the cost assessment studies. Fankhauser's estimates are also pegged to the $2*CO_2$ equivalent concentration, and tend to support the Nordhaus damage range of 0.25 to 2% of GNP in developed countries.

Damage estimates for the less developed countries (LDCs) are about twice as high. The IPCC (1996) Working Group III reviewed new evidence and arrived at estimates of the same order of magnitude.

Costs

The process initiated by the Framework Convention on Climate Change (FCCC) in 1992 and reconfirmed by the Berlin Mandate in 1995 moved greenhouse warming to a respectable position on the international environmental policy agenda. Despite the increasing number of studies, estimates of the global costs of possible alternative international agreements still cover a wide range due to unresolved questions, such as how these agreements might reshape energy production and consumption, or overall economic development globally and in major world regions.

The debate revolves around two main issues. Some model calculations show little long-term effect on atmospheric GHG concentration from rush and aggressive emission reductions. Others argue that at least the low or negative cost options for CO_2 abatement should be used, and initial price signals given to markets and technological development about the possible need for more ambitious emission reductions in the future. The second major issue is still open despite a number of plausible explanations: if studies identifying negative tails of the cost curves are correct, why are these opportunities to save money and CO_2 emissions not used?

When comparing a diverse set of models including optimization, regression-based, optimal growth and applied general equilibrium (AGE) models, it is not surprising that estimates of the costs of achieving an emissions target relative to the 1990 level of CO_2 emissions by some future date are sensitive to the reference case emissions trajectory projected by the individual model. Models with a higher reference case projection of CO_2 emissions will require more adjustments to reach any fixed targets than models with a lower reference case emissions projection.

Another typical observation from model comparison studies is that even when GDP growth rates are standardized, a very wide range of reference case emissions projections are produced. Projections of CO_2 emissions by the year 2100 range all the way from a 20% to a 200% increase over 1990 levels in various models. Relatively small differences in model parameters lead to large differences when their effects are compounded over the study's 110-year time horizon. For example, much of the difference in projections from the models for 2100 can be explained by differences in the assumed rate of autonomous energy efficiency improvement (AEEI) independent of energy price changes. The Global 2100 model (Manne and Richels, 1992) uses a value of 0.5% per year for this parameter, while the Edmonds-Reilly model (Edmonds and Reilly, 1985) assumes 1.0% per year. The more disaggregate assumptions made in the Global Macro model (Pepper, 1988) imply a rate of about 1.25%. When compounded over 110 years, these differences produce very different projections of aggregate energy use and emissions.

2.2. Integrated cost/benefit studies

The first attempt to bring together a diverse set of benefit and cost assessments into a consistent cost-benefit framework was by Cline (1992). The strong asymmetry between the two terms of the cost-benefit ratio reflect the imbalances in the number of studies and reliability of results of the global warming studies then available.

Cline identified 16 damage categories. He surveyed a relatively large number of studies to make his own damage estimate for each category. Due to the large number of source studies and the wide diversity of assumptions behind them, it is a major task to derive damage estimates that fit together. On the cost side, Cline made an in-depth survey of six state-of-the-art global energy-economy models and developed a synthesis based on their results. Damage assessments and cost estimates are then synthesized in a cost-benefit model. It incorporates other aspects of climate policy, such as costs of reduced deforestation and increased afforestation. Final results are then presented as 36 combinations of the four key parameters: the social rate of time preference, the climate sensitivity factor, the exponent of the damage function, and the base value of the damage function. Cline's conclusion from his cost-benefit analysis is that the "benefits of an aggressive program of abatement warrant the costs of reducing the GHG emissions if policy-makers are risk averse, or if one is pessimistic and concentrates on high-damage cases" (Cline, 1992, p. 311).

It is interesting to note that this conclusion is in sharp contrast to the other early cost-benefit assessment of climate change by Nordhaus (1991) who found that the optimal policy to reduce emissions of GHGs would entail a drastic cutback in CFCs. Carbon emissions should only be reduced by 2% and an optimal carbon tax of US$ 7.33 per ton would be sufficient.

2.3. Towards integrated climate-economy models

Studies conducted before serious attempts were made towards integrated assessments, dealing with the economic aspects of global climate change, have produced major developments in several areas. A variety of new ideas and new results has emerged. An incomplete list of new developments includes the following:

- Many features of the global warming problem make traditional methods of analysis difficult to apply, or inadequate. New ideas and innovative approaches are in great demand in order to make our economic, social, and technological analyses of climate change more relevant for policy makers. The approach to estimating agricultural impacts based on Ricardian rents by Mendelsohn et al. (1993) or the technological bifurcation analysis by Hourcade (1993) are excellent examples of the kinds of creative thinking necessary to overcome barriers of traditional analytical tools.

- There has been a gradual increase in the geographical coverage of damage estimates. This was made possible by the proliferating regional and national climate impact assessments conducted in many world regions. Although the methodological underpinnings of these studies are, at best, mixed, and many of them do not permit us to derive monetary estimates, we now have a substantially improved knowledge base for damage assessments than for the initial attempts, which applied a simple multiplier to derive damage estimates for LDCs from those calculated for MDCs.
- The time horizon of the analysis has been dramatically extended. Economists have traditionally considered time horizons of 20 to 30 years at most. The very long-term nature of climate change demands analyses on much longer time scales. Recent analyses face this challenge: Cline's (1992) analysis covers 300 years, some of Nordhaus' analyses with the DICE model extend over 400 years. These time scales, of course, raise new problems especially for the parameters affecting the intertemporal allocation of resources, notably the discount rate.
- Parallel to the increasing time horizons, there is a clear tendency away from the comparative static analyses based on $2*CO_2$ equivalent impact and damage assessments towards truly dynamic analyses. Various types of dynamic energy-economy models have been used to prepare cost estimates for many years, but dynamic approaches have only recently been applied in the benefit calculations.

All these helped open the way towards a new line of research. Models and analyses in this field are the topic for the next two sections of this paper.

3. Integrated assessments: Simulation models

Simulation models are intended to analyse a broad range of alternative futures. This implies changes in model parameters as well as user-defined scenarios. Simulation models usually do not contain a single criterion to determine the optimal outcome. A set of system indicators is reported at the end of each run and it is up to the model user to compare results. Developments in software technology make it relatively easy to modify scenarios by adjusting one or more model parameters but special care is needed to keep the selected scenario parameters consistent.

3.1. Fully integrated models

In studies of global warming, fully integrated models are those that cover the full cycle of GHG emissions through the biogeochemical processes to the impacts of climate change. For practical and computational reasons, modelers apply drastic simplifications when describing these complex systems. Perhaps the most dramatic condensing affects the representation of the atmosphere-climate system. Lessons from highly complex global circulation models are summarized in a few simple equations.

Socio-economic systems do not get a much better treatment either. The world economy is often represented by a single region or it is broken down into a few major regions of special interest to the modelers. Economic activities are described by a production function in which a composite commodity representing all goods and services is produced by labor, capital and energy. Reductions in carbon emissions are achieved by reducing energy availability in the models. This leads to a series of adjustments, representation of which depends on the model type, but the final outcome is some decline in total output. This output loss is considered to be the economic cost of CO_2 abatement. Costs related to curtailing emissions of other GHGs are calculated using the same principle.

One example of the fully integrated simulation models is PAGE, the model for Policy Analysis of the Greenhouse Effect (Hope et al., 1993). It is a probabilistic model that includes a simple representation of all important elements of climate change from emission policies and control costs to impact mitigation strategies and damages. To demonstrate effect of individual perceptions of the problem, the model uses multi-attribute utility functions.

3.2. Partially integrated models

As discussed in Section 2.1 above, there are models that link a few components of the anthropo-genically-induced climate change but do not cover the full cycle from emissions to impacts. Some of these models still offer useful insights into specific aspects of the problem.

One of the first integrated models was IMAGE, the Integrated Model to Assess the Greenhouse Effect (see Rotmans 1990; Rotmans et al., 1990). Over the past few years, the model was taken by several research groups. It was integrated into other models (see, for example Hulme et al., 1995), and it was also further developed to include new components. The version briefly described here is IMAGE 2.0, a multidisciplinary, integrated simulation model covering the global socio-economic, climate, and biosphere systems (see Alcamo, 1994).

I classified the current version of IMAGE 2.0 as partially integrated because available reports do not indicate either the costs of emission control measures or the approaches to climate impact assessment and their economically measurable implications. This version of IMAGE includes three modules: Energy-Industry, Terrestrial Environment, and Atmosphere-Ocean. The model has a global coverage and a time horizon to the year 2100, with a calibration period of 1970 to 1990.

Spatial resolution varies across modules from large global geopolitical regions in the socio-economic model, down to a 0.5° latitude by 0.5° longitude resolution in the vegetation model.

The group of simulation models to which IMAGE 2.0 also belongs permits experimentation with a very broad range of scenarios. This might be both a curse and a blessing. On the one hand, it is great to have a tool that permits the implications of unconventional or extreme future scenarios, like the population of the globe becoming vegetarian or the global energy demand being fulfilled from biomass energy, to be explored. Yet there is little empirical base to re-estimate model parameters for cases that are very different from the original estimation period. Retuning all affected parameters in a consistent way is no mean task, if possible at all.

4. Integrated assessments: Optimization models

4.1. Fully integrated models

Similar to our classification scheme for simulation models, I also categorize optimization models into the two major groups defined above: fully integrated models and partially integrated models. When it comes to formulating optimal policies for GHG abatement, it is important to know whether the advice comes from a globally aggregated model or from one that has at least some regional detail. Finally, while models aggregating the total economic activity into a single production function provide useful insights as a first approximation, we are also interested in the intersectoral implications of the various strategies within individual regional or national economies. However, these models pose major challenges in their conceptual formulation and raise major difficulties when one tries to load them with reliable data.

Global single-region, aggregated models

The most compact representation of the optimal climate policy problem is provided by models that treat the globe as one region, the world economy as a single representative producer-consumer. GHG emissions are computed from the level of activity in the sector producing one composite good. Both costs of emission abatement and damages from GHG-induced climate change are fed back to deteriorate the production of total output.

The pioneering work in this field of climate economics is DICE, the Dynamic Integrated model of Climate and the Economy by Nordhaus (1994). Given the extremely long time lags between GHG emissions and their economic impacts, the concept and models of optimal growth offer a convenient framework for analysis. Nordhaus took the model version formulated by Frank Ramsey in 1928 and extended it to include both impacts of anthropogenic climate change and the allocation of resources to reduce those emissions.

The optimization criterion in DICE is to maximize the discounted sum of the utility of *per capita* consumption, that is the value of a traditional social welfare function. Paths of optimal growth affected by climate change are diverted from the unconstrained optimal path as a result of losses in output due to global warming and diverting resources to reduce emissions. Similar single-region, sectorally aggregated global models were formulated by Fankhauser (1994) and Maddison (1995).

Global multi-region, aggregated models

There are profound differences in the level of development, economic structure, energy use and associated carbon emissions between countries of the world. This suggests that even if a globally optimal climate policy could be defined, its implementation would need to take into account these regional differences. Uniform reduction in CO_2 emissions worldwide thus might not be the best solution for political and economic reasons. Effects of a global tradable CO_2 emission permit scheme on the abatement side and differentiated regional damages on the impact side can be analysed only by regionally disaggregated models.

One example of this type of model is MERGE, the Model for Evaluating Regional and Global Effects of GHG reduction policies (see Manne et al., 1995). It consists of three modules: a new version of the Global 2100 model (see Manne and Richels, 1992), called Global 2200, which is a fully integrated applied general equilibrium model with look-ahead dynamics. The second module deals with changes in the atmospheric concentrations of GHGs and the temperature changes induced by them.

The third module contains the assessment of damages from global warming. In the market damage category, MERGE adopts the assumption that damages rise quadratically with temperature change. MERGE is one of the first models that attempts to include nonmarket impacts explicitly in the damage assessment. To get a crude assessment of this damage category, the willingness-to-pay approach is taken. The basic question is how much consumers in each region would be willing to pay to avoid ecological damage. The two main factors affecting the outcome are predicted temperature change and GDP *per capita*.

Sectorally disaggregated models

Disaggregated models in this classification are those in which the economic module describes the economy with more than one sector and accounts for production and consumption processes of more than one economic good. In addition, the model should also include a simple representation of the climatic processes and at least rough damage estimates. No such model is currently available, although precursors can be detected.

Jorgenson and Wilcoxen (1993) calculate costs of various CO_2 abatement strategies. By estimating parameters of a highly disaggregated (I would call it "top-to-deep-down") general equilibrium model econometrically from long historical data sets, the authors give their model a respectable memory of long-term evolution processes. This makes all model parameters and especially elasticities more suitable for long-term future analyses than single-point parameterization. Although the perfect substitution assumption used in the model does not permit modeling the depletion of fossil fuel sources, this is not an important limitation as proven geological stocks will not be depleted over the model's time horizon of roughly one century. This powerful tool is then used to evaluate macroeconomic costs of different GHG policy instruments for the US economy.

Jorgenson and Wilcoxen compare the costs of a carbon tax, a tax on the energy content of fossil fuels (a BTU – British thermal units – tax), and an *ad valorem* tax on fuel use. Their results suggest that the least expensive solution (carbon tax) has a drastic effect on one sector (coal mining), while the most merciful tax for coal miners (the *ad valorem* tax) has a much greater effect and implies much higher costs for the economy as a whole.

Closing the loop both at the atmosphere/climate side (which will require estimates of non-US CO_2 and global non-CO_2 GHG emissions) and at the optimal resource allocation side, similarly to DICE, will by no means be a straightforward task, but it is not difficult to imagine that it will be done soon. The result will be a powerful tool for integrated cost/benefit assessments, at least for the American economy.

4.2. Partially integrated assessments

Uncertainties are profound in each step of our analysis as we go from emissions to radiative forcing to predicted climate change to impact assessment. Uncertainties tend to accumulate in integrated assessments producing results where uncertainty ranges around markedly different scenarios having a substantial overlap. These results provide limited insights into the problem as they are difficult to interpret from a policy perspective.

One response to these difficulties might be that analysts turn back and limit the scope of their work to the most essential components of the integrated framework. Probably the best recent example of this kind of analysis is that of Richels and Edmonds (1995).

The concept of critical load turned over the acid rain debate and served as a useful tool to guide policies to abate emissions contributing to the acid deposition problem. Similarly, the global warming debate could be considerably simplified if we had a target established for atmospheric concentrations of GHGs beyond which induced climate change would reach intolerable scales. Even in the absence of such well-established targets, analyses of the relationships between hypothetical concentration limits, the amount and temporal path of emissions, and the associated costs of control provide useful information, especially for short-term climate policy. These analyses help identify critical points in the decision process, e.g. "point of no return" in emissions when delayed action cannot keep peak concentrations within a given target, no matter how ambitious it is.

Richels and Edmonds combine two global energy models and a reduced form carbon cycle model in their assessment. The energy models are the Edmonds-Reilly-Barns (ERB) partial equilibrium model and the Global 2100 dynamic nonlinear optimization model. Alternative emission paths for achieving any prespecified level of atmospheric concentrations are calculated by the impulse-response function of Maier-Reimer and Hasselmann (1987). Although the level of integration in this case is rather low (emissions, concentrations, costs), the Richels-Edmonds analysis provides interesting lessons about the costs of various short-term policies and their long-term impacts on concentrations.

5. Layout for the book

Papers in this volume are organized into two blocks. Following this background essay, papers in Part 1 cover important conceptual issues. They are supplemented by a succinct report from the insurance industry. Part 2 begins by looking at GHG reduction targets currently discussed and includes two contributions analysing the potential role of present and future energy technologies. Final papers in the volume incorporate observations on the international policy process and target-based assessments following up on the contribution to the First Conference of the Parties held in Berlin by the scientific panel advising the Federal Government of Germany on global environmental issues.

A central issue in the analyses of efficient environmental regulation over the past two decades has been whether to control prices to correct an environmental externality or quantities of pollutants allowed to be emitted. Gary Yohe takes a fresh look at this old dilemma in the context of climate policy. The insights from his paper suggest an important policy implication. Designers of international agreements and related policy instruments would be well advised to consider the distribution of efficiency gains that can be harvested under various schemes over the shorter and longer term to make them really attractive to all parties.

Due attention to the long time scales involved in the climate change problem in analysing policies and instruments is also advised in the paper by Jürgen Blank. He provides a short summary of merits and shortcomings of the most important instruments currently included in energy-economic models. Searching for incentives for nation states to participate in international agreements on GHG emission reductions, he identifies the interdependence of costs and benefits of mitigation policies. A proper treatment of these issue, he suggests, requires integrated assessment models operating at the national or at least regional scales.

As we learned more about potential impacts of climate change on various ecosystems and economic sectors, it has become increasingly clear that the most serious risks loom outside the national accounts. Forests and other types of terrestrial vegetation, coastal and other aquatic ecosystems have limited capacity to adapt to changing climate, especially if the rate of change is unprecedentedly fast. Possibilities for managing and helping them adapt are largely unexplored and the associated costs also unknown. Agriculture, forestry and water management were the first market sectors subjected to thorough impact assessments. The paper by Gerhard Berz starts out with a summary of past weather-related damages and their consequences for the insurance industry. These issues are then taken up in the context of future climate-related risks as perceived by analysts in the branch.

Over the past years, the debate about climate policy proceeded in two major directions. One course focused on what can be considered as the ultimate task in managing the risk of climate change, namely stabilizing the atmosphere. As formulated in Article 2 of FCCC, this entails stabilizing atmospheric concentrations of all GHGs at levels necessary to avoid "dangerous anthropogenic interference with the climate system".

The first steps in the direction of these long-term objectives, nonetheless, involve short-term mitigation objectives and measures to reach them. Numerous short-term objectives have been informally declared or formally approved so far. Eberhard Jochem evaluates the European stabilization targets both in terms of the expected emission reductions and the related costs. The paper by Peter Hennicke complements Jochem's analysis by taking a closer look at the energy sector, currently available technologies, and mitigation options.

Jae Edmonds and his co-authors also address energy technologies as a key issue in managing global warming. In contrast to the two preceding papers, however, they adopt a long-term view. If the ultimate objective is to stabilize atmospheric concentrations of GHGs rather than emissions, this offers much more flexibility in timing and locating abatement efforts according to when and where they are cheapest. The costs of achieving the same atmospheric target could be drastically reduced by giving proper signals to markets and allowing time for the development of low-cost low-carbon or carbon-free technologies. While the cost-saving potentials associated with new energy technologies are clearly demonstrated by the models adopted by Edmonds et al., we know much less about the "proper" magnitudes of the necessary market signals and the true dynamics of technological development that are expected to produce them.

Cost estimates are probably the single most important inputs analysts provide to the policy process. Short-term costs are important because climate policies need to compete for funds with numerous other pressing social, economic, and environmental problems. Long-term costs are important mainly in the context of potential climatic risks related to delayed emission reductions. While modelers are far from being able to give unequivocal answers to all aspects of the cost questions, if indeed it will ever be possible, the international policy process picks up its own momentum and proceeds towards a climate protocol.

It is impossible for anything other than a newspaper to provide an up-to-date account on the policy process. Michael Ernst summarizes current events and expectations in the policy process. His contribution provides an excellent basis for comparing earlier plans and objectives for the international negotiations with actual achievements, say, after the Third Conference of the Parties in Kyoto in December 1997.

The challenges involved in managing the Earth's atmosphere have triggered various activities at the science-policy interface. Modeling exercises, review and assessment activities, and advisory panels provide information for the policy process at the national and international level. In addition to the global-scale IPCC process, many governments established their own science committees to provide advice.

The German Advisory Council on Global Change (WBGU) prepared a special report in 1995 on the occasion of the First COP held in Berlin (WBGU, 1995). The report involves a simple target-based inverse analysis. Maximum values for absolute global mean temperature and for its rate of change are derived from the Earth's geological history to define a so-called "tolerable climate window". A compact climate and carbon cycle model is then used to derive acceptable emission paths that keep the climate system within those predefined limits. While the Richels-Edmonds study mentioned in Section 4.2 analyses alternative paths to reach predefined concentration targets, the WBGU study goes one step further and works with (admittedly simple) climate targets.

This volume contains two follow-ups to the WBGU analysis. Issig and his co-authors attempt to incorporate atmospheric aerosols into the WBGU modeling framework. Although their linear approach to this complex problem characterized by highly nonlinear processes is rather crude, the basic message appears plausible. Due to the masking effect of aerosols, temperature windows based on observed temperatures are likely to be oversized compared to what the temperature record should be without masking. Nevertheless, this point becomes really relevant at the regional scale because of the well-known regional differences in sulphur emissions.

The contribution by Petschel-Held and Schellnhuber also takes the WBGU model as its starting point. The authors provide an overview of a novel approach to integrated target-based climate modeling, the so-called Tolerable Windows Approach. This approach extends the WBGU analysis into a fully-fledged integrated assessment. The approach ranges from tolerable impacts of climate change on selected exposure units to help define the climate window on the impact side, to developing cost-effective strategies to reach those targets on the abatement side. The

TWA now serves as a conceptual basis for a major international research project coordinated by PIK.

6. Closing remarks

Even a small workshop and its proceedings convey a broad range of opinions on the scientific and policy issues involved in climate change. These often conflicting views were represented at the workshop and have also been preserved in this collection. It is important to bear these differences in mind when trying to derive policy guidelines from the various models and assessments. Besides, ironing them out would in any case have been impossible.

Given the enormous uncertainties in our current knowledge about the climate system and implications of its change for ecosystems and society, probably the best strategy is to let a thousand flowers bloom. Diversity of IAMs also helps answer the heterogeneous set of strategic questions raised by a broad range of policy makers. Efforts to compare IAMs (like the Energy Modeling Forum) and critically appraise them (like the IPCC process) as well as fora to discuss specific issues involved in their development and possible uses (like the present workshop) will continue to be important for future improvements of these models both in terms of their scientific quality and policy relevance.

References

Alcamo, J (ed.) (1994) IMAGE 2. *Integrated modeling of global climate change*. Kluwer, Dordrecht.

Ausubel, JH (1991) Does climate still matter? *Nature* 350:649–652.

Cline, WR (1992) *The economics of global warming*. Institute for International Economics, Washington, DC.

Edmonds, J and Reilly, J (1985) *Global energy: assessing the future*. Oxford University Press, New York.

Fankhauser, S (1993) The economic costs of global warming: Some monetary estimates. *In*: Y Kaya, N Nakičenovič, WD Nordhaus and FL Toth (eds): *Costs, impacts, and benefits of CO$_2$ mitigation*. IIASA, Laxenburg.

Fankhauser, S (1994) The social costs of greenhouse gas emissions: An expected value approach. *Energy Journal* 15(2):157–184.

Hope, C, Anderson, J and Wenman, P (1993) Policy analysis of the greenhouse effect. An application of the PAGE model. *Energy Policy* 21(3):327–338.

Hourcade, JC (1993) New challenges for energy-environment long-term modeling: Lessons from the French case. *In*: Y Kaya, N Nakičenovič, WD Nordhaus and FL Toth (eds): *Costs, impacts, and benefits of CO$_2$ mitigation*. IIASA, Laxenburg.

Hulme, M, Raper, SCB and Wigley, TML (1995) An integrated framework to address climate change (ESCAPE) and further developments of the global and regional climate modules (MAGICC). *Energy Policy* 23(4/5):347–355.

IPCC (1996) *Climate Change 1995. Economic and Social Dimensions of Climate Change*. Cambridge University Press, Cambridge, UK.

Jorgenson, DW and Wilcoxen, PJ (1993) Reducing US carbon dioxide emissions: An assessment of different instruments. *In*: Y Kaya, N Nakičenovič, WD Nordhaus and FL Toth (eds): *Costs, impacts, and benefits of CO$_2$ mitigation*. IIASA, Laxenburg.

Kaya, Y, Nakičenovič, N, Nordhaus, WD and Toth, FL (eds) (1993) *Costs, impacts, and benefits of CO$_2$ mitigation*. IIASA, Laxenburg.

Maddison, D (1995) A cost-benefit analysis of slowing climate change. *Energy Policy* 23(4/5):337–346.

Maier-Reimer, E. and Hasselmann, K (1987) Transport and storage of carbon dioxide in the ocean – an inorganic ocean circulation carbon cycle model. *Clim. Dynam.* 2:63–90.

Manne, AS and Richels, RG (1992) *Buying greenhouse insurance – the economic costs of CO$_2$ emission limits.* MIT Press, Cambridge.

Manne, AS, Mendelsohn, R and Richels, RG (1993) MERGE – A model for evaluating regional and global effects of GHG reduction policies. *Energy Policy* 23(1):17–34.

Mendelsohn, R, Nordhaus, WD and Shaw, D (1993) The impact of climate on agriculture: A Ricardian approach. *In*: Y Kaya, N Nakičenovič, WD Nordhaus and FL Toth (eds): *Costs, impacts, and benefits of CO$_2$ mitigation.* IIASA, Laxenburg.

Nordhaus, WD (1991) To slow or not to slow: The economics of the greenhouse effect. *Econ. J.* 101(6):920–937.

Nordhaus, WD (1994) *Managing the global commons: The economics of climate change.* MIT Press, Cambridge.

Pepper, W (1988) Global macro-energy model. Summary paper. ICF Incorporated, Fairfax.

Richels, RG and Edmonds, J (1995) The economics of stabilizing atmospheric CO$_2$ concentrations. *Energy Policy* 23(3/4):373–379.

Rotmans, J (1990) *IMAGE: an integrated model to assess the greenhouse effect.* Kluwer, Dordrecht.

Rotmans, J, de Boois, H and Swart, RJ (1990) An integrated model for the assessment of the greenhouse effect. *Climatic Change* 16:331–356.

Schelling, TC (1992) Some economics of global warming. *Am. Econ. Rev.* 82(1):1–14.

WBGU (German Advisory Council on Global Change) (1995) *Scenario for the derivation of global CO$_2$ reduction targets and implementation strategies.* WBGU, Bremerhaven.

First principles and the economic comparison of regulatory alternatives in global change[1]

Gary W. Yohe

Department of Economics, Wesleyan University, Middletown, CT-06459, USA

1. Introduction

Comparisons of the relative economic efficacy of alternative regulatory strategies have been available for more than two decades (e.g. Poole, 1969; Weitzman, 1974; Laffont, 1977; Yohe, 1978). The results are well established in the economics literature, and their import has propagated into many subdisciplines – the economics of regulation (e.g. Karp and Yohe, 1977), international trade (Pelcovits, 1976; Stiglitz and Dasgupta, 1977), macroeconomic policy (Poole, 1969) and environmental and resource economics (Spence and Roberts, 1976; Yohe, 1979, 1992) to name a few. The results are, nonetheless, frequently misrepresented and misapplied even within those literatures; and so it is safe to say that the intuition that underlies their content is not thoroughly understood.

Notwithstanding practical and political differences and complications derived most directly from issues of the implementation and administration of various regulatory schemes, it therefore seems worthwhile to review that intuition yet again in the isolated and artificial realm of the first principles of economic thought. This is not to say that economic reasoning must carry the day when alternative strategies are considered and that regulatory schemes designed to maximize efficiency, even in an elaborate economic model, must be implemented. It is, instead, to say that intuition built from first principles can help to identify clearly those circumstances in which the efficiency consequences of a regulatory design choice could be large and others in which they are likely to be small; and so the intuition will help to identify circumstances in which the "What 'value' economic efficiency?" debate should be conducted and others in which it is not worth the effort.

---

[1] This paper is a revised version of an earlier paper published in the June 1997 issue of the *OPEC Review*. It appears here with permission of the publishers.

2. Taxes versus standards versus marketable permits under uncertainty

The key to understanding the intuition of regulatory comparisons under uncertainty is to understand that each regulatory strategy allows its own signature pattern of variability whose net value may be positive or negative. In the context of pollution most generally and greenhouse gases more specifically, for example, it is essential to understand that variability (in emissions) is a knife that cuts two ways. On the one hand, variability is a reflection of economic actors' abilities to respond to changes in their circumstances even in the face of some sort of environmental regulation. These responses must be efficient and improve both the pace of economic activity and the direct welfare supported by that activity; otherwise, they would not be undertaken and neither economic activity nor social welfare would deviate from expected and anticipated levels. The point is that economic actors can always choose not to vary their behavior as their circumstances change even though the control mechanism that they face allows them to do so; when they chose to do so, therefore, it must be true that they, at least, are made better off.

On the other hand, the public goods character of pollution problems means that variability in emissions produces corresponding variation in their associated social cost. As a result, variability increases the expected cost of emissions to a degree that is indicated by the convexity of the estimated social cost schedule. The selection between standards, taxes and permits is thus a task of balancing the additional private benefits accruing from whatever degree of variability in economic activity and emissions that each structure might allow against the additional social costs that this variability will impose.

Figure 1 replicates the standard representation of the Weitzman (1974) model highlighted and exploited in Karp and Yohe (1977). It is a construction designed to facilitate the comparison of the two extreme regulatory options – prices (carbon taxes, e.g. in global warming context) and standards (fixed targets without joint activity of any kind); in the climate change context, indeed, Figure 1 is the portrait of any integrated assessment that builds on the cost-benefit paradigm. Schedules (B) and (C) drawn there represent expected marginal benefit and marginal cost curves for carbon emissions. The marginal benefit curve reflects, of course, the derived demand for carbon – a demand drawn from the positively valued goods and services whose fabrication and/or delivery involve (e.g.) the combustion of fossil fuel. The marginal cost curve similarly reflects damages cum adaptation costs associated with emissions through increased concentrations and the physical impacts that result. Note that the marginal benefit curve calibrates private benefits, for the most part, while the marginal cost curve displays social damages.

Schedules (B) and (C) can support a comparison between a tax that maximizes expected net welfare with its certainty equivalent standard; this is the Weitzman comparison between a tax set at T and a standard set at Q, for which there would be no efficiency grounds for selecting one of

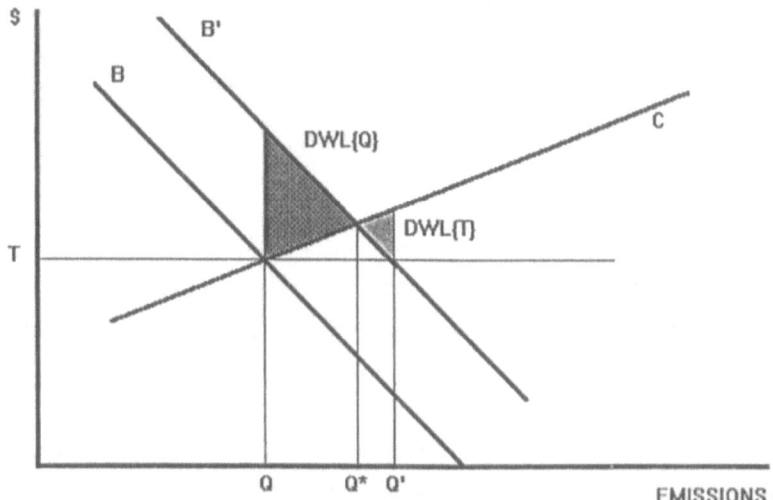

Figure 1. Designation of deadweight loss for a tax [DWL{T}] and the certainty equivalent quantity standard [DWL{Q}] for the circumstance in which marginal benefit is higher than expected.

the other if (B) and (C) were both known with certainty. If the precise location of either or both schedules were unknown before the choice had to be made, however, things change. To see why, consider a state of accelerated economic activity that would shift the derived demand for carbon up (by an amount η). The result would, under an assumption that the regulators' views of expected benefits and damages would not change so quickly, mean that schedule B'(η) would operate against either the tax T or the standard Q. Under the tax, emissions would rise to Q'(η); fixed by the standard, however, emissions would remain at Q. The Weitzman insight is that neither of these outcomes is economically efficient given B'(η). Indeed, Q*(η) would be maximally efficient given B'(η); and so both the tax and the standard impose some deadweight welfare loss upon the system.

The area designated DWL{Q} is the deadweight loss associated with the standard; it represents private gains in excess of additional social cost that would have been generated if emissions had be allowed to climb to Q*(η). The area DWL{T} is the deadweight loss associated with the tax; it represents additional social cost in excess of private gains that would be incurred as emissions climbed from Q*(η) up to Q'(η). A comparison of T and Q over all possible states of economic activity can be accomplished by computing what Weitzman would call, in this case, the comparative advantage of the tax over the standard:

$$\Delta_{TvsQ} = E\{DWL(T) - DWL(Q)\}.$$
$$= (0.5)[b-c]Var\{Emiss\} \qquad [1]$$

In writing Equation [1], –b and c represent the slopes of the derived demand (marginal benefit) and marginal cost curves, respectively, and Var{Emiss} indicates the variance of emissions allowed under the tax alternative across all possible "states of nature" (see Weitzman, 1974 or Yohe, 1978 for a derivation). If Δ_{TvsQ} is positive (negative), then the deadweight loss associated with the standard is, on average, greater (smaller) than the loss associated with tax and the tax (standard) should be preferred.

Equation [1] clearly depends upon the linear structure of Figure 1, and so its applicability to a general comparison of the tax and standard alternatives is limited. Its content, though, informed by the geometry of the figure and the recognition that variation in emission is neither all good nor all bad, can support the first principles intuition toward which this paper is directed. Table 1 records a summary of the insight that can be drawn from both, particularly if one looks at the extreme values that critical parameters might assume. Equation [1] does, however, reveal here that those parameters include the slopes or elasticities of both the derived demand for carbon emissions and their associated marginal social damage schedule. The variance of carbon emissions also matters in determining the importance of making the right choice (the size of Δ_{TvsQ} but not its sign); but that effect is determined, in part, by the elasticity of derived demand.

Notice that Table 1 can support a qualitative taxonomy of supporters for taxes and strict standards. Researchers who are concerned about surprises and/or nonlinearities in damages even

Table 1. Comparative advantage of taxes over standards – a qualitative taxonomy

$$D_{TvsQ} = (.5)[b-c]Var\{Emiss\}$$

Case	Limiting conditions	Explanation	Preference	Global change context
(1) Large c	$c \to \infty \Rightarrow \Delta \to -\infty$	Variation significantly increases expected costs	(-----) STANDARD	Nonlinearities and surprises on the damage side
(2) Small c	$c \to 0 \Rightarrow \Delta \to (+)$	Variation produces small increase in expected costs	(+++) TAXES	Linear damages along smooth trajectories with adequate adaptation
(3) Large b	$b \to \infty \Rightarrow \Delta \to 0^+$	Very little variation even with the tax option	(+) TAXES	Derived demand for carbon emissions very insensitive to price changes
(4) Small b	$b \to 0 \Rightarrow \Delta \to -\infty$	Very large variation under the tax regime	(-----) STANDARD	Derived demand for carbon emissions very sensitive to price changes
(5) b=c	$b=c \Rightarrow \Delta \to 0$	Deadweight losses match	0 INDIFFERENCE	Identical sensitivities so that the deadweight losses are equal, but not necessarily small

Table 2. Comparative advantage of taxes over permits – a qualitative taxonomy

$$D_{TvsPer} = (.5)[(.5)b-c]Var\{Emiss\}$$

Case	Limiting conditions	Explanation	Preference	Global change context
(1) Large c	$c \to \infty \Rightarrow \Delta \to -\infty$	Variation significantly increases expected costs	(-----) PERMITS	Nonlinearities and surprises on the damage side
(2) Small c	$c \to 0 \Rightarrow \Delta \to (+)$	Variation produces small increase in expected costs	(+++) TAXES	Linear damages along smooth trajectories with adequate adaptation
(3) Large b	$b \to \infty \Rightarrow \Delta \to 0^+$	Very little variation even with the tax option	(+) TAXES	Derived demand for carbon emissions very insensitive to price changes
(4) Small b	$b \to 0 \Rightarrow \Delta \to -\infty$	Very large variation under the tax regime	(-----) PERMITS	Derived demand for carbon emissions very sensitive to price changes
(5) b=c	$b=c \Rightarrow \Delta \to +$	Deadweight losses do not match	(+) PERMITS	Permits allow some efficient variation between firms that fixed standards do not

associated with low probability events should prefer the stability in outcome offered by standards. People who are concerned about wide variability in emissions even around a well-established mean should do the same. Others, though, who see global change as a gradual and relatively predictable process and envision wide options for timely adaptation should do the opposite – forsake the stability of strict standards for the efficiency gains associated with the maximal flexibility allowed by taxes.

Similar analyses of expected deadweight losses can compare taxes with marketable permits and permits with strict standards – comparisons that are particularly important when the sources of the derived demand for carbon emissions are diverse and widely distributed. Table 2 summarizes the results of the taxes/permits comparison based upon an analogous comparative advantage calculation for prices versus permits. It is a comparison whose algebra ends with:[1]

$$\Delta_{TvsPer} = E\{DWL(T)-DWL(Per)\}.$$
$$= (0.5)[(0.5)b-c]Var\{Emiss\}. \qquad [2]$$

Table 2 tends to the extremes of Equation [2], and so it mimics Table 1 to a significant degree. The key to understanding its content again lies in tracking the sources and consequences of variability in emissions. Careful reading of Equation [2] and Table 2 reveals that the comparison

---

[1] The algebra that supports equation (2) can be found in Appendix A of Yohe (1992).

of taxes and permits still weighs the efficiency of the maximal variability in emissions allowed by taxes against the stability of a permit scheme. The difference, now, is that permits allow maximal variability in emissions across the various sources of derived demand *subject to the constraint that total emissions are held fixed.* Some efficient variation in sources' emissions is allowed by permits, but fixing total emissions eliminates any associated increase in expected damage.

The taxonomy that emerged from Table 1 therefore still applies, but the deck is stacked slightly more in favor of permits than strict standards. Notice, for example, that the economic value of even the limited variability allowed by permits is enough to turn the indifference of the fifth case in Table 1 (where $b=c$) into a clear preference toward permits and against the tax alternative. Put in a different context, the comparison between standards and permits is easy to dismiss. Compared with standards that fix emissions at every source, permits increase efficiency without any increase in expected damage; and so permits must, on economic grounds at least, always be preferred. A corollary is that any close economically-based comparison of taxes and standards suggests that marketable permits, working as a structural compromise between two extremes, should probably be preferred to either extreme.

3. Marketable permits versus joint implementation under certainty

It is becoming increasingly clear that creating an international marketing system for emissions permits in which all nations can participate, and encouraging joint implementation between smaller groups of countries or individual emitters in individual countries, are two entirely different and distinct policies. They may, in theory, both produce the same pattern of emissions, but the distribution of the efficiency gains that they support can be dramatically divergent. Figure 2 provides some simple structure within which to make this point; it can also be used to suggest that the dimension of the distributional divergence can be quite large.

Schedule B_I in Figure 2 represents the derived demand for carbon emissions from a country or group of countries whose emissions are initially to be constrained. It is convenient to think that B_I represents the marginal benefit to be derived from Annex I countries in the context of the FCCC and the Berlin Mandate. Their demand is measured against the left side vertical axis so that emissions are measured left to right from origin O. In the absence of any restrictions, Figure 2 shows that these countries would choose to emit an amount equal to Q_{Imax}. Other countries' derived demand (non-Annex I countries referred to as A_{II} countries in the Figures and denoted by the subscript "II" in the text and equations) is reflected by schedule B_{II} drawn against the right hand vertical axis, with emissions being measured right to left from origin E. In

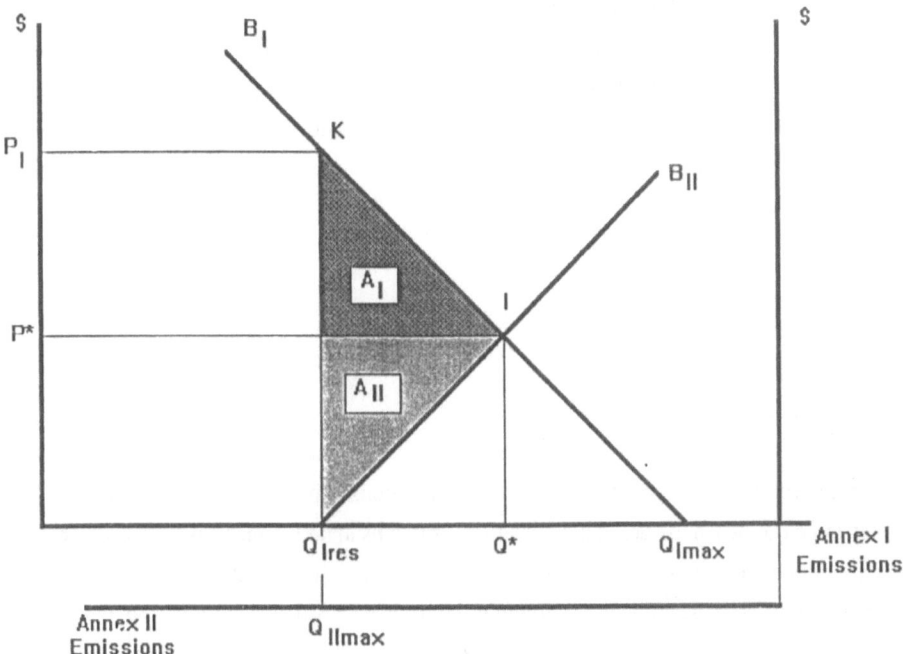

Figure 2. Allocation of emissions and consumer surplus between Annex I and non-Annex I countries under marketable permits when an emissions standard is imposed only on Annex I countries.

the absence of any restrictions, these countries would choose to emit an amount equal to Q_{IImax}. Together, then, both groups of countries would emit a total amount equal to $\{Q_{Imax}+Q_{IImax}\}$.

The Berlin Mandate holds, however, that the emissions of Annex I countries will be restricted, and the location of origin E relative to origin O reflects a required reduction in their emissions equal to amount $\{Q_{Imax}-Q_{Ires}\}$. Total emissions are, in this construction, therefore equal to distance OE with Q_{Ires} allocated initially to Annex I countries and Q_{IImax} allocated to the members of non-Annex I. The shadow price of Q_{IImax} in non-Annex I is zero, in this allocation; but the shadow price is P_I in Annex I. The difference between P_I and zero highlights the potential for efficiency gain; all that is required is some means of allowing Annex I countries to shift their emissions reduction responsibilities to non-Annex I until the shadow prices are equalized.

Marketable permits would, of course, do the trick. Schedule B_I, particularly below point K, represents a demand curve for permits for Annex I countries. Similarly, schedule B_{II} is an offer curve (i.e. a supply curve) for permits that non-Annex I countries would be willing to sell. The market would clear at P^*, and the sale of permits would effect the transfer of $\{Q^*-Q_{Ires}\}$ units of emission from non-Annex I to Annex I. Area $\{A_I+A_{II}\}$ captures the dimension of the efficiency gain and suggests its distribution. Of that area, the market would allocate "consumer surplus" equal to area A_I to Annex I countries and "producer surplus" equal to area A_{II} to non-Annex I countries. Letting the slopes of the derived demand curves be given by $(-b_I)$ and $(-b_{II})$ for Annex I and non-Annex I countries, respectively, and representing Annex I emissions reduction responsibilities by $e = \{Q_{Imax}-Q_{Ires}\}$, some simple algebra shows that the total efficiency area is equal to:

$$\{A_I + A_{II}\} = (e^2 b_I^2)/2(b_I + b_{II})$$

of which

$$A_{II} = (b_{II} b_I^2 e^2)/2(b_I + b_{II})^2$$

is allocated to non-Annex I countries.

This share is small (indeed, the potential efficiency gain is small, as well) when b_I is small (i.e. when the derived demand for emissions in Annex I nations is relatively elastic) and when b_{II} is relatively large (i.e. when the derived demand for emissions in non-Annex I is relatively inelastic). The share allocated to non-Annex I nations can also be small when their derived demand for emissions is elastic (i.e. when b_{II} is small) but the story is a bit different. In this case, the potential for efficiency gain can be large even though little of its value flows to the supply side of the market. Finally, elastic demand in Annex I (large b_I) produces the opposite story – the share flowing to non-Annex I can be a large fraction of a large total.

Reality is likely to lie somewhere within the boundaries of these extremes, of course, so the question of allocation can be significant. Suppose, for the sake of simplicity, that the two derived demand schedules had the same slope; i.e. assume that $b = b_I = b_{II}$. The share allocated to non-Annex I would then be 50% of the total:

$$A_{II} = be^2/8$$

Assuming that non-Annex I nations would not face emissions restrictions for a long period of time and that unregulated Annex I emissions would grow annually at some rate g, the discounted value of the non-Annex I share can be approximated simply as

$$PV\{A_{II}\} = be_o^2/8(r-2\ g) \qquad\qquad [3]$$

where e_o represents current Annex I emissions and r is a discount rate.

Non-Annex I nations will not escape restrictions forever, however; and the simple analytics of Figure 2 must change when that happens. Figure 3 provides a portrait of this situation, with Annex I still required to reduce emissions by $\{Q_{Imax}-Q_{Ires}\} = e$ and non-Annex I now required to reduce emissions by $\{Q_{IImax}-Q_{IIres}\} = e_{II}$. Notice that area A_c now represents cost incurred in non-Annex I. Note, as well, that the efficiency gain from trading is now a smaller area, but that the

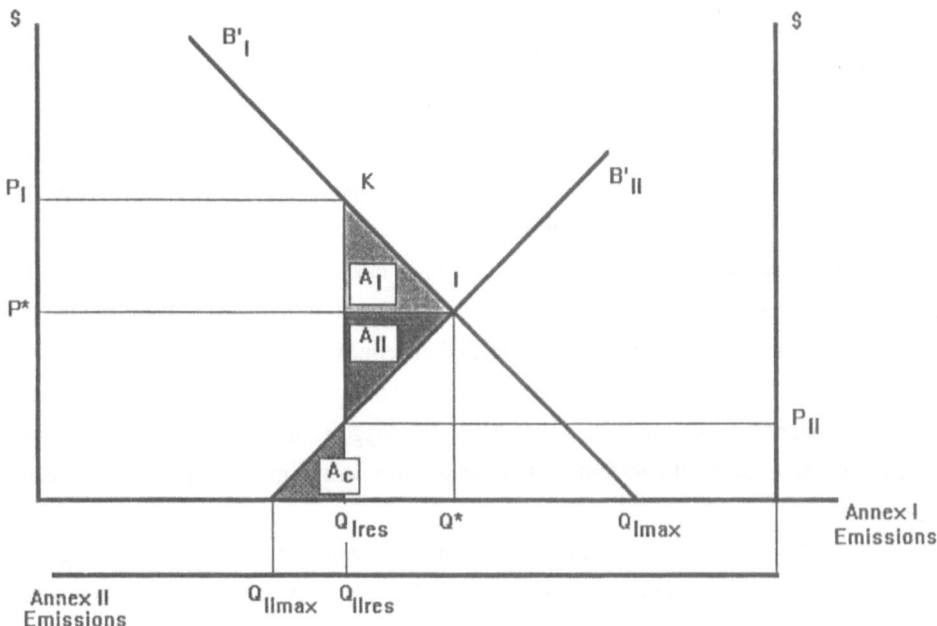

Figure 3. Allocation of emissions and consumer surplus between Annex I and non-Annex I countries under marketable permits when emissions restrictions are imposed on all countries.

market still allocates area A_{II} to non-Annex I. The net position of non-Annex I nations is now $\{A_{II}-A_c\}$. Assuming, again, equal slopes, this position is given by

$$\{A_{II}-A_c\} = b(e-e_{II}-4e_{II}{}^2)/8.$$

Letting T represent the length of time that the non-Annex I nations will be free of any emissions restrictions and representing the rate of growth of required emissions reductions by g_I and g_{II} for Annex I and non-Annex I, respectively, the discounted value of this net position from then on can be expressed as

$$PV\{A_{II}-A_c\} = bexp\{-r(T)\}\{[e_o/(r-g_I)]-[e_{II}/(r-g_{II})]-[4e_{II}/(r-2g_{II})]\}/8 \qquad [4]$$

The precise formulation of this value is less important than the observation that it need not be positive. Indeed, in the likely event that $g_I<g_{II}$ at the time when restrictions hit non-Annex I countries, it is quite likely that $PV\{A_{II}-A_c\}$ will be negative because non-Annex I costs would increase as their source of revenue diminishes.

The key to non-Annex I participation may therefore be the careful collection of their share of the initial efficiency gain before they face the cost of reducing their own emissions. Taking some liberty with the data and the structure for purposes of exposition, application of the MERGE model of Manne and Richels to an initial EMF-14 modeling exercise suggested that the share of global carbon emissions produced by OECD countries was something like 50% in 1995, that those emissions might grow by something like 1.3% per year to, say, 2025, and then by 0.4% per year thereafter.[2] The rest of the world's share might correspondingly be expected to be double the OECD 1995 level by 2025, but their emissions could grow by as much as 1.7% per year from then until, say, 2080. Plugging these suggested "data" into a formula that computes the ratio of revenue generated in non-Annex I by the sale of permits to 2025 ($PV\{A_{II}\}$ from Equation [3]) to the net cost of meeting an emissions restriction defined by 2025 emissions ($PV\{A_{II} - A_c\}$ from Equation [4]) produces a quotient roughly equal to unity. This is an artificial manipulation of one particular "what if" scenario, to be sure, but it suggests the possibility that the present value of the entire stream of non-Annex I transactions in a permits market could actually net out to zero.

Interestingly enough, a quotient equal to unity also comes close to matching the ratio computed from the more careful but nonetheless preliminary analysis conducted by Manne and Richels of a similar scheme that fixes OECD emissions at 80% of 1990 levels by 2010 with non-OECD countries' emissions eventually limited by actual 2020 levels.[3] There, allocating 50% of the cost savings from marketing permits from 1995 to non-OECD countries (perilously assuming constant slopes of their respective aggregate-derived demand schedules) produces roughly $750 billion in revenue (in present value) – an amount that seems to match the approximate present value of the net cost of achieving the 2020 emissions stabilization downstream.

It is in pondering the distribution of these sums of money ($750 billion in present value is more than 3.5% of current world GNP) that a comparison of permits with joint implementation has a critical point to make. Joint implementation allows, in fact encourages, bilateral and other limited arrangements between Annex I nations and/or their citizens and non-Annex I nations and/or their citizens. As such, joint implementation runs the risk of losing the "invisible hand impersonality" of a market. There is, more to the point, no reason to believe that Annex I negotiators will not, when operating in a limited context with a few "suppliers of emissions reduction transfers", be able to extract all of the efficiency gains reflected not only by area A_I in Figures 2 and 3 (as is their due in the marketable permit scheme), but also area A_{II}. Appealing to the standard theory of perfect price discrimination, in fact, an argument can be made that making one

---

[2] See Standard Reference Case figures for emissions in the EMF-14 Summary of Results from November 23, 1995.
[3] See Figures 7 and 8 in Manne and Richels (1995).

deal at a time would allow Annex I countries to move deliberately up the non-Annex I offer curve – accepting deals in which the compensation sent to non-Annex I would exactly cover marginal cost and nothing more. The efficient solution Q^* would again obtain, but all of the potential producer and consumer surplus would be expropriated by Annex I.

This is the same sort of result that would be obtained if a profit maximizing monopolist were able to extract (or extort, depending upon your point of view) from each customer the maximum price that he or she would·be willing to pay. The monopolist would, in that case, continue to produce and sell product as long as the marginal cost of production were no greater than the lowest price that the last customer would be willing to pay; and since that price would reflect the marginal benefit of the last unit purchased, the monopolist would produce up to the output for which marginal cost exactly matched marginal benefit – i.e. the optimal output. All of the consumer surplus would be collected by the monopolist at the expense of the consuming public, of course, but at least the level of production would be "correct" in an efficiency sense.

Cast in the context of the long-term prospects for non-Annex I, this sort of exploitation would rob them of perhaps all of the positive terms in the present value computation of their international emissions position. Cast in terms of their participation in any sort of international control mechanism, therefore, the news is bleak. Why should non-Annex I nations participate in joint implementation even if it might generate huge efficiency gains for the globe, if they reap none of those gains? The answer is, of course, that there is no reason for them to do that. But there is also no reason that non-Annex I nations could not create their own market mechanism. They need not negotiate on a limited basis. They could set up their own market for joint implementation contracts and sell them at whatever price the market will bear. All they need is the wherewithal (1) to identify and to quantify emissions reduction opportunities that Annex I nations would find attractive (i.e. identify the projects reflected by the lower region of their derived demand curve in Figures 2 and 3), (2) to assess honestly and accurately the competitive market value for those opportunities (i.e. compute the points on their derived demand curve highlighted above in comparison with the corresponding region of demand from Annex I), and (3) to have the courage to walk away from any deal that does not reflect P^* to some degree. Put simply, then, marketable permit schemes require some initial allocation of emissions rights; and it is important to note that the Berlin Mandate characterizes one such allocation.

4. Concluding remarks

The economic analysis presented here is very "low-tech". It is rooted in the fundamental concepts with which economists assess economic efficiency – supply and demand, producer and consumer surplus, the distribution of income, and the public goods problem of the global commons. It applies directly, therefore, only in the rarefied air of an economic construction; and then really only to the parts that are linear at least in approximation. Nonetheless, the intuitive results are applicable beyond the constructions cited here.

First of all, first principles show clearly that variation in emissions must have a good side. Emissions would not vary otherwise. But first principles show that variation in emissions has a bad side, as well; and therein lies the comparative rub. When contemplating alternative regulatory strategies in all of their comparative complication, therefore, there is an efficiency tradeoff that needs to be part of the discussion.

Second, market power, reflected here as an ability to dictate price and quantity, is important in determining the distribution of efficiency gains. Marketing permits produce those gains, and competitive markets produce a distribution in which all parties share in their surplus at least to some degree. There may be disagreement over whether that distribution is appropriate or desirable, but at least it appears virtually by itself. Joint implementation could, by way of contrast, play into the "market power" of large nations and distort a competitive distribution dramatically – perhaps to the point where smaller nations would decide not to participate. Given that the recent and preliminary analysis of these potential efficiency gains suggests that they could be enormous, first principles warn that inattention to the details of distributional patterns could be very costly.

References

Karp, G and Yohe, G (1977) The optimal linear alternative to price and quantity controls in the multifirm case. *J. Comp. Econ.*3:56–68.

Laffont, J (1977) More on prices vs. quantities. *Rev. Econ. Stud.* 44:177–182.

Manne, A and Richels, R (1995) The Berlin Mandate: the Costs of Meeting Post-2000 Targets and Timetables, presented at the National Forum on Global Climate Change in Columbus, Ohio on September 7.

Pelcovits, M (1976) Tariffs vs quotas. *J. Int. Econ.* 6:363–375.

Poole, W (1969) Optimal choice of monetary instruments in a simple stochastic macro model. Special Studies Paper Number 2. Washington: Federal Reserve Board.

Spence, M and Roberts, M (1976) Effluent charges and licenses under uncertainty. *J. Public Econ.* 5:193–207.

Stiglitz, J and Dasgupta P (1977) Tariffs vs. quotas as revenue raising devices under uncertainty. *Am. Econ. Rev.* 67:975–981.

Weitzman, M (1974) Prices vs quantities. *Rev. Econ. Stud.* 41:50–65.

Yohe, G (1978) Towards a general comparison of price controls and quantity controls under uncertainty. *Rev. Econ. Stud.* 45:229–238.

Yohe, G (1979) Managing the demand side of a sudden supply shortage. *Resources and Energy* 2:221–241.

Yohe, G (1992) Carbon emissions taxes: their comparative advantage under uncertainty. *Ann. Rev. Energ. Environ.* 17:301–26.

Cost-Benefit Analyses of Climate Change: The Broader Perspective
F.L. Toth (ed.)
© 1998 Birkhäuser Verlag Basel/Switzerland

Dynamics of policy instruments and the willingness to participate in an international agreement

Jürgen E. Blank

Westfälische Wilhelms-Universität Münster, Lehrstuhl für Volkswirtschaftstheorie, Universitätsstr. 14–16, D-48143 Münster, Germany

1. Introduction

The greenhouse problem can be characterized by the following features:

- The damage is not caused by the flow of emissions but merely by the global accumulated stock in the atmosphere. Therefore neither the region nor the time of emissions (within a range of some decades) is relevant for the damage. This gives the theoretical advantage, that a carbon emission tax or some kind of certificate are reasonable instruments for environmental policy. The optimal time path of global emissions must be determined in such a way that the greenhouse gas concentration does not exceed a critical value.
- Not only must the stock restriction for greenhouse gas concentration in the atmosphere be considered; the exhaustibility of fossil fuel stocks, especially of oil and natural gas, is essential for the determination of the extraction time path, and hence the emission path of carbon.
- On the other hand, this peculiarity creates all possible problems of incentives of international cooperation: The individual free-rider position both *vis-à-vis* other regions or future generations is a severe impediment for any environmental policy. Furthermore, the costs of mitigating greenhouse gas emissions in any country depends on the response of other countries.
- Since the damage is caused by the stock of greenhouse gas emissions we have very long time horizons for the calculation of damages caused by greenhouse gases.

This raises three further problems: First, since all these future damages can only be calculated by means of some long-term simulation models, there is a high degree of uncertainty stemming from the unavoidable complexity of the models and the large number of very uncertain parameters. Today we have reliable information only on a certain range of substitution possibilities, technologies, demand reactions, etc., but the long-term adjustments will probably go far beyond that range of known parameters.

The second consequence of the very long-term structure of the problem is to the economic evaluation of costs of the reduction of greenhouse gas emissions today (whose costs have to be carried by people in the next decades) and the benefits in the form of avoided catastrophes or other losses far in the future. In other words, the discount rate and the degree of risk aversion *vis à vis* potential catastrophic events determine any good cost-benefit argumentation.

The third problem is the degree of confidence in the quality of modeling: Whereas in many other areas, modeling can be understood as one helpful method among others for a better understanding of complex problems, there is in fact no alternative to modeling with respect to the very long-run cost-benefit considerations.

Modeling the energy-economic module in computable models of global warming, however, still has a lot of shortcomings, especially in recognizing the dynamics of the greenhouse effect and hence the dynamics of policy instruments to mitigate greenhouse gases (GHG). Furthermore, the interdependencies of costs and benefits of mitigation policies and the willingness of nations to participate in an international agreement to reduce greenhouse gas emissions should also be stressed. In particular, the behavior of fossil fuel suppliers, acting strategically if confronted by the threat of any measure to reduce greenhouse gases, is poorly recognized in existing computable energy-economic models (Blank and Ströbele, 1995).

2. Impacts of different kinds of policy instruments

Existing energy-economic models typically take a form of international agreement among all countries, aggregated to several regions, to cut back emissions by some percentage compared with a specific base year (Toronto type agreement). Each region chooses a given policy instrument (mainly emission taxes) to reach this target. It is well known that this does not give an efficient outcome. The reasons are twofold. First, the same global reduction can be achieved at lower costs through a different distribution of regional emission reductions. Second, since the climate is affected through cumulative emissions, an exogenously-given reduction path is not cost minimizing. If damages are caused by cumulative emissions rather than by current emissions the date when a certain unit is emitted is less relevant.

The energy-economic models consider different types of policy instruments (but largely taxes) to mitigate the problem of global warming but stipulate passive behavior of all agents. For example, the strategic and dynamic reactions of fossil fuel suppliers are neglected in nearly all of the global energy-economic models, like Global 2100, GREEN, and the Edmonds-Reilly-Barns (ERB) model.

In general, the range of policy instruments used to reduce the problem of global warming is quite large.[1] The world models, typically using the top-down approach, have focused on a tax on

---

[1] An overview is given in the Second Assessment Report of the IPCC (1996), Working group III, chapter

the carbon content of fossil fuel consumption. Less attention is given to tradable permits. In these world models, it is assumed that any region chooses its policy instruments, in general this is the tax rate on the consumption of fossil fuels, in such a way that an exogenously given emission path is met. This is only one kind of a tax regime. Alternatively, taxing energy consumption or a combination of energy/carbon taxation is also possible. Another kind of taxation would be to tax the production of carbon or energy and other greenhouse gases. Whatever kind of taxation scheme is chosen, from the point of view of a benevolent UN-planner a cost-effective solution is easy to implement. But the distributional effects are quite different. The distributional effects are essential for a country if it decides to join an international agreement or not. Due to the exhaustibility of fossil fuels, each producing country gets a so-called hotelling or scarcity rent. Any measure to reduce the consumption of fossil fuels will have a rent-extracting effect, to the burden of the producer. To reach a great number of participants in an international agreement, some form of side payments would be necessary. Some computable world models allow for trade in emissions rights to capture side payments in their models.

Since the greenhouse problem is due to the level of concentration of greenhouse gases in the atmosphere, the problem of determining the optimal global emission path is not trivial. One of the problems that should be solved is how fast the concentration ceiling should be reached. Energy-economic models cannot give the answer to this specific problem. Integrated assessment models, connecting the energy-economic models together with physical and climate models, are necessary. Until now, the evaluation of different kinds of policy instruments has been done only under the consideration of a given emission path.

Due to the complexity of modeling the global warming problem, most energy-economic computable equilibrium models are static. One exception is Global 2100/2200 by Manne/Richels, which is a dynamic optimization model, but has shortcomings in modeling international trade and shows only five geopolitical regions. But since the greenhouse gas problem is a stock- and not a flow-restriction, static modeling yields misleading results. It is already a very heavy task to model dynamic behavior in an interdependent world[2], but incorporating international negotiations into computable world models seems nearly impossible.

A variety of policy instruments to control greenhouse gas emissions could be applied on multitude levels (see Fig. 1). In general, a tax regime or a system of tradable emissions rights could be implemented on the global as well as on the individual level. But whatever level is chosen, the aim of any policy instrument is to influence the behavior at the individual level, i.e. of households and firms, which affect the atmosphere by their economic activities. At this level, individual actors decide how much they would reduce their greenhouse gas emissions or, alter-

---

[2] A first approach was done by Fisher et al. (1994).

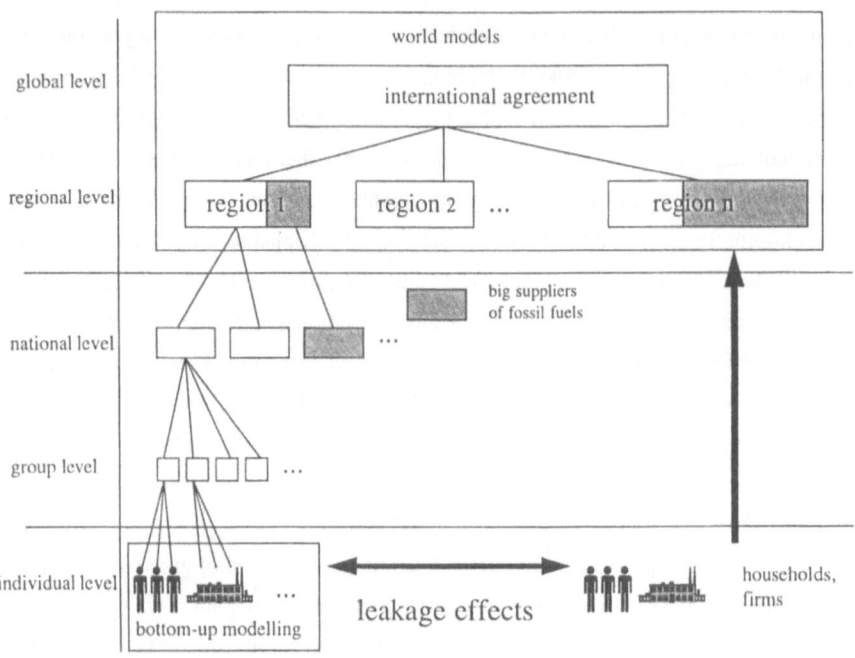

Figure 1. Institutional levels for policy instruments.

natively, look for ways of avoiding the costs of any policy instrument. A unilateral imposition of policy instruments on the national level in particular gives rise to leakage effects. GHG-emitting industries will move where environmental restrictions are less costly. Furthermore, some countries might gain by attracting such kinds of industries. For example, fossil fuel-producing countries might invest in energy-intensive industries and would become large exporters of energy-intensive goods. The extent of the leakage effect depends on the number of participating countries in an international agreement to mitigate greenhouse gas emissions, and the kind of policy instrument chosen. The problem of how policy instruments affect the international trade pattern, and therefore give rise to leakage effects and hence affect the willingness to participate in an international agreement, should be recognized and integrated in energy-economic models.

In the following paragraphs some policy instruments will be briefly discussed with respect to the willingness to join an international agreement to mitigate greenhouse gases.

2.1 Taxes

The basic idea of using a tax as a policy instrument to mitigate current and future emissions of greenhouse gases is to raise the price in such a way that the demand for greenhouse gas-emitting products will meet a desired emission path. In the case of a worldwide planner maximizing a utility function, an optimal tax scheme can be derived easily. In this case, it is insignificant whether the tax is imposed on the consumption or the production of greenhouse gases. But if we have individual nations who have to decide what an optimal tax regime should look like, then the question of whether the tax should be levied on the consumption or on the production of greenhouse gases (or fossil fuels as the main source of greenhouse gases) is of importance. In the former case, the tax revenue would be collected by consuming countries, in the latter the tax would be collected by fossil fuel-producing countries. The distributional effects are completely different. In the case of an excise tax the burden falls totally on the producer countries as long as the supply side is modeled as a perfect competitive market. Hence, market supply prices have to decline compared to a situation without any taxation. Suppose OPEC were to internalize a carbon tax at the wellhead, would the consumer countries join such an agreement? On the other side are all models imposing taxes on the consumption of fossil fuels.[3] Why should OPEC participate in such agreement? The price for fossil fuels would decrease drastically. OPEC could attract energy-intensive industries from participating countries if it stays outside of an agreement.

This problem might be avoided if the tax were to be imposed on both the consumption and the production of greenhouse gases or, alternatively, the total international tax revenue, collected by some international institution, were to be shared among consuming and producing countries according to some redistribution rule. However, this would make an international agreement much more complicated. Therefore, a worldwide tax regime, covering all countries in an international agreement, seems very unrealistic.

The decision makers have to set the tax rate (independent of whether the tax were to be imposed on consumption or production, independent of whether carbon or energy were taxed) considering all the reactions of energy consumers over a period covering the next 150–200 years. This would be neither politically (lack of credibility) and technically (no knowledge of the production function in 2100) feasible. This argument is valid whether the tax is implemented on a global or a national level. Therefore, the tax rates have to be adjusted from time to time, with all the problems of renegotiation.

Furthermore, a tax applied at the global level reduces the range of national policy instruments. If a tax is implemented on a global level, individual countries cannot choose other instruments.

---

[3] One exception is Whalley and Wigle (1991), who analyze the effects of a tax imposed on the production of fossil fuels.

Alternatively, instead of a global carbon tax, a domestic tax may be implemented in two ways. First, each country chooses its own tax rate to maximize its net benefits. In this case, tax rates would vary across countries due to the different consequences of global warming for different countries. Besides the problem of free riding, some countries may act strategically in such a way to attract polluting industries from abroad and benefit from leakage effects. Second, a domestic tax is chosen to reach a given emission path. It is well known that this is not cost effective (either intertemporally or internationally) due to the nontradable quota character. Joint implementation may be seen as an instrument to enforce cost effectiveness. An alternative is to let the quota be tradable.

2.2 Tradable permits

A tradable permit scheme imposed on the international level allows individual nations to choose among the total range of domestic policy instruments. Trading permits leads to a cost effective allocation of emissions between countries, but the costs (price for emissions rights) are uncertain.

Incentives to participate in an international agreement to mitigate greenhouse gas emissions depend on the initial quota allocation scheme.[4] Grandfathering and GDP-weighted emissions are advantageous to the industrialized world, whose greenhouse gas emissions have already reached very high levels. As it is disadvantageous for industrializing countries, whose greenhouse gas emissions will rise drastically (China and Far East), these countries are unlikely to participate. If such a scheme is implemented, greenhouse gas emissions will decline over the next decade, but in the long run one has to expect rising emissions from the nonparticipating countries. A scheme of equal *per capita* emissions shows the opposite effect: Industrialized countries are more injured, due to high *per capita* emissions, than industrializing countries with low *per capita* emissions. A high free-lunch option will lower the costs for buying emission rights. Using Joint Implementation results in lower costs in the short run, but might result in higher costs in the long-term, due to the neglect of investment in less carbon-intensive production technology.

Therefore, economists should consider how initial quota allocation regimes should be designed to reach an international agreement in which all individual nations would participate. To be cost effective, permits must be tradable not only among nations/individuals but also over the whole time period. Note that for the concentration level it is insignificant whether a unit of carbon

---

[4] In the case of carbon, one can have tradable permits on the use or on the production of fossil fuels, resulting in different distribution schemes and, in the case of limited participation in an international climate agreement, they have different economic consequences (see Hoel, 1992). Whether permits are on the use or on the production of fossil fuels, there is a rent-extracting effect in both cases.

is emitted today or in 50 years' time. There is no institutional framework to guarantee the validity of such kinds of permits. Therefore, the optimal emission path must be chosen in the first period; thereafter permits valid for short periods can be granted. An adjustment of the global emission path in the course of time is possible. The cost effectiveness of the chosen emission path depends on the number of participants in an international agreement.

2.3 Joint implementation

Joint implementation is seen to reduce the cost of greenhouse gas emission in industrialized countries. Instead of reducing emissions in their own country, industrialized countries invest in greenhouse gas emission-reduction measures in other – mainly developing – countries. The idea behind this is cost effectiveness. To reduce a given amount of greenhouse gas emissions in an industrialized country costs much more than the same amount of reduction in a developing country. But joint implementation is not a policy instrument designed for a global solution to global warming. It might be a first step towards implementing an international tradable permit system. But if such a system is established, market forces would ensure cost effectiveness. There will be no need for joint implementation.

3. Participation in an international agreement to mitigate greenhouse gases emissions

The aim of an international agreement to mitigate the emission of greenhouse gases is to achieve a particular concentration target in such a way that cost effectiveness is reached. Such a target can be achieved in several ways. Figure 2 shows three alternative individual greenhouse gases emission paths, A, B, and C.

The area under each emission curve is equivalent to the same net accumulation of greenhouse gases up to time T, which might be some 150 years ahead. The aim is to identify that emission path that minimizes the costs of reaching the desired concentration level. If a country has great potential for no-regret policies, a path like C should be chosen. In case of only small no-regret potential and capital stock rigidities, time path A seems more appropriate. The same is true for global emission paths; a particular concentration target may be achieved by different time paths of emissions, some more costly than others. In the existing computable world energy-economic models dealing with accumulated carbon restrictions, the global emission path is found by aggregating the regional emission paths. Mitigation measures would be realized in that region where it is cheapest to do so. But this approach ignores the problem of burden sharing. Due to

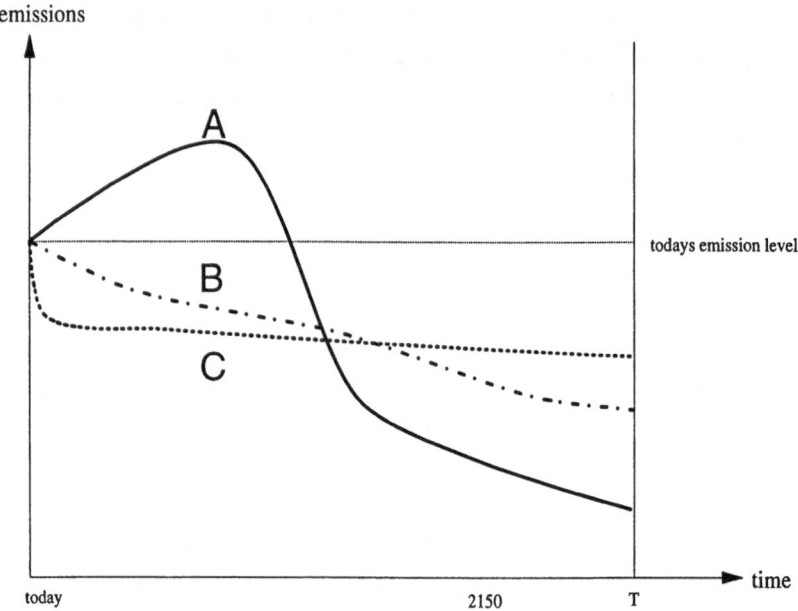

Figure 2. Alternative emission paths under the same greenhouse gases concentration.

burden sharing, it is much more complicated for a set of countries to select a global cost-effective emission path than for a worldwide planner. Individual countries or regions are interested in how the damages of global warming and the mitigation costs are distributed among them. This might affect the willingness to participate in an international agreement.

Due to the global dimension of the greenhouse effect, all greenhouse gas-emitting countries should be participants in an international agreement. Even if all countries gained by reducing global warming, it would be very difficult to reach a global agreement. But if some countries gained and others lost, the implementation of an international agreement would become an enormous task. It is necessary to explore the cost and effectiveness of different kinds of international agreements to reduce greenhouse gases.

A first approach is taken by Edmonds et al. (1995). Using the Edmonds/Reilly/Barns model, they analyze the effects of a uniform carbon tax, tradable permits and individual national targets (non-tradable permits) on the regional costs and benefits. Due to the static structure of the ERB model, it is assumed that participants in an international agreement agree to hold emissions constant at rates equal to those at the time of initial participation.

As seen above, there is a variety of different emission paths to achieving a desired concentration level. But it might be that a certain emission path, like A, favors countries i_A and will cost countries i_C more than they benefit from greenhouse gas abatement, because these countries would prefer emission path C.

The number of participating countries in an international agreement depends on the amount of GHG reduction over a given time period. The solid line in Figure 3 shows the number of participants in an international agreement as a function of GHG reduction. As long as there is a free lunch option, it is plausible to expect that all nations or regions will join such an agreement, independent of the number of participating countries. But if the amount of greenhouse gas reduction is rising, it will become more costly. Nations whose discounted stream of costs is higher compared to the discounted stream of profits will not participate. If leakage effects are regarded, costs will rise even more, and hence the number of potential participants decrease (dotted line in Fig. 3).

Modeling cooperation among strategic players requires the concept of cooperative game theory. The payoff each participant will get in such an agreement depends on the costs and benefits. Economic-energy models are adequate to derive costs for GHG-reducing measures. We need integrated assessment models disaggregated on national or at least on regional levels to derive the benefits from any GHG reduction scheme.

Costs and benefits result from two sources: First, each country has to bear the direct costs of abatement and receive the benefits from the global GHG reduction. Second, costs and benefits are due to the interdependence of policy responses among countries. For example, any climate policy will affect the prices for fossil fuels and hence the price for energy intensive goods. Consumer

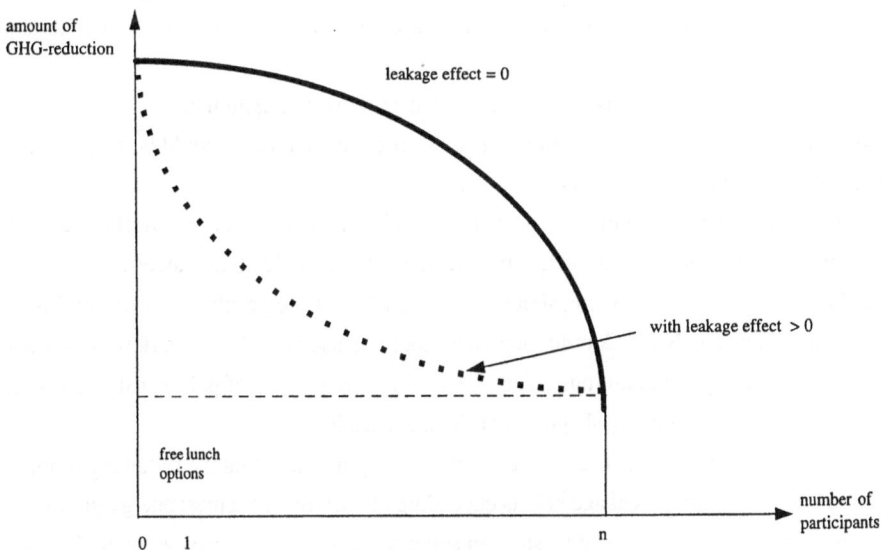

Figure 3. Amount of GHG reduction and the number of participants in an international agreement.

prices will rise, whereas the behavior of supply prices depends on the market structure for fossil fuels. In the case of a competitive fossil fuel market, fossil fuel prices will decrease. Nonparticipating countries might raise their greenhouse gases emissions due to the production of energy intensive goods. This is the well-known problem of leakage effects. Further, OPEC member countries will suffer from a decrease in oil prices, which negatively affects their willingness to participate in an international agreement. In the case of an oligopolistic fossil fuel market, with more market power for OPEC, fossil fuel prices might rise as a result of internalizing the greenhouse effect at the wellhead. In this case, any GHG reduction measures are more costly for countries with no stocks of fossil fuels. Hence, the impact of GHG reduction measures on the demand for fossil fuels is crucial to the market structure on the supply side.

4. Need for future research activities

The economics of the global warming problem is much more complicated than is recognized in the computable energy-economic models available today. New approaches mention the greenhouse gas problem as a capacity problem for the atmosphere, but the exhaustibility of the resource stock together with strategic fossil fuel supplier is still neglected.

The economic analysis of global warming policy has not taken adequate notice of the dynamics of the greenhouse gases problem, and hence the dynamics of policy instruments. Another weakness in the economic analysis is inadequate consideration of the interdependence of costs and benefits of policy responses among countries (see Fisher et al., 1994). Both problems should be treated jointly.

The new generation of models will have to meet the following requirements:

- Since the problem of global warming is a stock problem, countries should be modeled as forward-looking dynamic optimizers.
- Due to the important problem of participation in an international agreement, leakage effects are essential. Hence, it is necessary to incorporate international trade in the models.
- Furthermore, nations or regions should be modeled as strategic players, maximizing some utility function. The Nash-Cournot outcome should be used as a benchmark for international cooperation or 'carbon coalitions'. The effects of different types of policy instruments on the participation in an international agreement should be analyzed.

Since this is an enormous task, economists should start by investigating the following problems:

- Empirical investigation: leakage effects depending on different schemes: energy prices including measures to reduce GHG emissions in some regions, energy prices without GHG reduction measures in others.

- Distributional effects of various greenhouse gas instruments on large suppliers of fossil fuels.
- Confrontation of a dynamic view of world energy markets (exploration and transformation process of "almost sure" reserves into producing sources) with changed incentives from a global warming policy.
- Market structure and competitive forces (competitive fringe with significantly higher cost than OPEC suppliers) in the oil and gas market *via-à-vis* a global warming policy.
- Distributional effects of various global warming policy instruments with different redistribution schemes at the regional or national level.
- Incentives to participate in an international agreement.

Acknowledgements
I acknowledge helpful suggestions and comments from Wolfgang Ströbele. The author gratefully acknowledges financial support of the Volkswagen-Stiftung in Hannover for the research project: Interaction of suppliers of fossil fuels *vis-à-vis* a policy to reduce CO_2 emissions.

References

Blank, J and Ströbele, W (1995) *The economics of the CO_2-problem: what about the supply side?* Discussion Paper, No. 147/95, University of Oldenburg.

Edmonds, J, Wise, M and Barns, DW (1995) Carbon coalitions. *Energy Policy*, 23(4): 309–335.

Fisher, BS, Hinchy, M and Henslow, K (1994) *A dynamic game approach to greenhouse policy. ABARE* Conference Paper 94.2, presented at the Tsukuba Workshop 1994.

Hoel, M (1992) *Tradable emission quotas for CO_2: Quotas on use of carbon or on production of carbon?* Working Paper 1992:1, Oslo: CICERO.

IPCC (Intergovernmental Panel on Climate Change) (1996) *Climate Change 1995: Economic and Social Dimensions of Climate Change.* Cambridge University Press, Cambridge.

Whalley, J and Wigle, R (1991) The international incidence of carbon taxes. *In*: R Dornbusch and J Poterba (eds): *Global Warming: Economic Policy Responses*. MIT Press, Cambridge.

Cost-Benefit Analyses of Climate Change: The Broader Perspective
F.L. Toth (ed.)
© 1998 Birkhäuser Verlag Basel/Switzerland

Global warming and the insurance industry

Gerhard A. Berz

Münchner Rückversicherung, Königinstrasse 107, D-80802 München, Germany

1. Introduction

In the last few decades, the international insurance industry has been confronted with a drastic increase in the scope and frequency of great natural disasters. The trend is primarily attributable to the continuing steady growth of world population and the increasing concentration of people and economic values in urban areas. An additional factor is the global migration of populations and industries into areas such as coastal regions, which are particularly exposed to natural hazards. The natural hazards themselves, on the other hand, are beginning to increase significantly in some areas.

In addition to the problems the insurance industry has with regard to pricing, capacity and loss reserves, the assessment of insured liabilities, preventive planning and the proper adjustment of catastrophe losses are gaining importance. The present problems will be dramatically aggravated if the greenhouse predictions come true. The increasing intensity of all convective processes in the atmosphere will force up the frequency and severity of tropical and extratropical cyclones, tornadoes, hailstorms, floods and storm surges in many parts of the world with serious consequences for all types of property insurance.

Rates will have to be raised, and in certain coastal areas insurance cover will only be available after considerable restrictions have been imposed, such as significant deductibles and low liability or loss limits. In areas of high insurance density, the loss potential of individual catastrophes could reach a level at which the national and international insurance industries will run into serious capacity problems. Recent disasters showed the disproportionately high participation of reinsurers in extreme disaster losses and the need for more risk transparency if the insurance industry is to fulfill its obligations in an increasingly hostile environment.

2. Increasing disaster losses

During the early 1990s, our knowledge of global and regional climatic processes increased considerably, not least thanks to financial support for climate research. Parallel to this, a remarkable series of major natural disasters taught the insurance industry bitter lessons it will not forget. Table 1 lists all natural disasters from the past few decades that have cost the insurance industry more than US$ 1 billion. While only one event, Hurricane Alicia, exceeded this figure prior to 1987, the period since 1987 has seen 16 such events, 14 of them since 1990. Of these, Hurricane Andrew is the clear leader with insured losses of some US$ 20 billion, which, however, would have been much higher had Andrew not landed a "double miss" but had achieved two direct hits on Miami and New Orleans. The same is true of the 1994 earthquake in California, which only skirted the edge of Greater Los Angeles and, in spite of insured losses of some US$ 12,5 billion, can only be considered a "warning shot".

The loss trend since 1960 (see Fig. 1) clearly shows the dramatic increase in catastrophe losses in the last few years – a development that could well see average annual loss burdens from great natural disasters rise to US$ 30–50 billion (in today's values) by the end of the decade. The increase compared to the 1960s, which in the 1980s amounted to a factor of three for economic losses and a factor of five for insured losses, has since escalated to factors of six and

Table 1. The billion dollar insurance losses from natural disasters*

Rank	Year	Event	Area	Insurance losses (mio. US$)	Economic losses (mio. US$)
15	1983	Hurricane "Alicia"	USA	1.275	1.650
6	1987	Winterstorm	Western Europe	3.100	3.700
5	1989	Hurricane "Hugo"	Caribbean, USA	4.500	9.000
4	1990	Winterstorm "Daria"	Europe	5.100	6.800
14	1990	Winterstorm "Herta"	Europe	1.300	1.900
9	1990	Winterstorm "Vivian"	Europe	2.100	3.250
13	1990	Winterstorm "Wiebke"	Europe	1.300	2.250
3	1991	Typhoon "Mireille"	Japan	5.200	6.000
11	1991	Forest fire	USA	1.700	2.000
1	1992	Hurricane "Andrew"	USA	20.000	30.000
12	1992	Hurricane "Iniki"	Hawaii	1.600	3.000
10	1993	Blizzard	USA	1.750	5.000
17	1993	Floods	USA	1.000	12.000
2	1994	Earthquake	USA	12.500	30.000
7	1995	Earthquake	Japan	2.900	100.000
16	1995	Hailstorm	USA	1.150	2.000
8	1995	Hurricane "Opal"	USA	2.100	4.000

*As of 11/95

Great Natural Disasters
1960 - 1995
as of October 1995

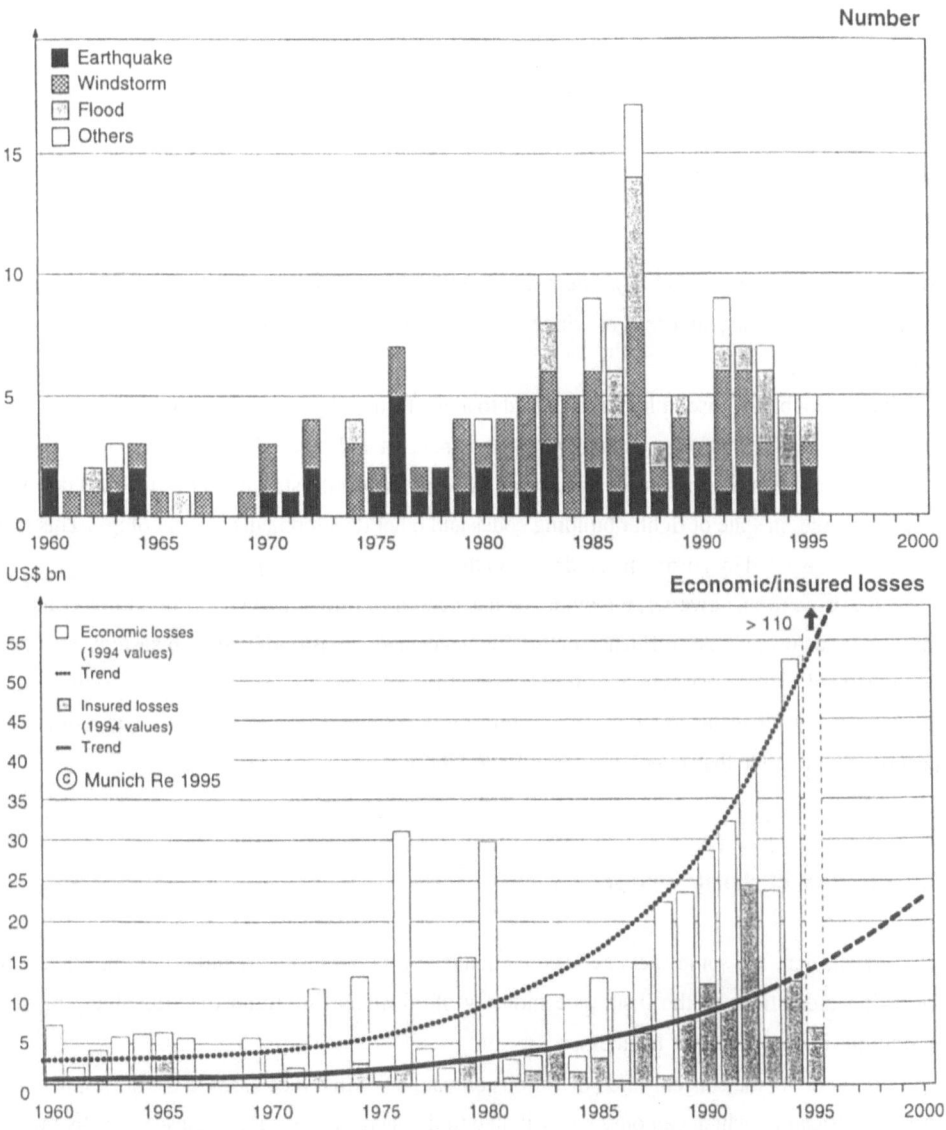

Figure 1. Great natural disasters (1960–1995).

Table 2. Great natural disasters 1960–1995

	Number of events	Economic losses*	Insured losses*
Decade 1960–1969	16	44.5	6.0
Decade 1970–1979	29	85.6	10.0
Decade 1980–1989	70	135.9	27.4
Las ten years 1986–1995	76	362.0	86.9
Factor			
80s: 60s	4.4	3.1	4.6
last ten: 60s	4.6	8.1	14.5

*Figures are in US$ billion (1994 values)

fourteen respectively (see Tab. 2). These data are based on "great" natural disasters. The remaining natural loss events, hundreds of which are recorded each year, at least double the overall loss volume.

Without doubt, this rise in losses is caused to a great extent by increasing economic values and insured liabilities, especially in heavily exposed areas. In addition, natural disasters have shown time and again that the loss susceptibility of buildings and infrastructures has increased rather than decreased, in spite of tighter building codes and other developments in technology. This fact was again illustrated by Hurricane Andrew and the Los Angeles earthquake.

At the same time, however, the evidence has become stronger that emerging climate change might have an ever greater influence on the frequency and intensity of natural disasters. This was demonstrated by the great windstorm disasters of recent years that set new loss records almost every year, as well as the countless flood, storm, drought and forest fire disasters that seem to be occurring more often than ever before (see Berz, 1993).

3. Indications of climate change

First of all, here are some facts about the changes in the atmosphere and on earth, which are worrying indications to the insurance industry of an increasing trend towards changing risk conditions in many areas of business.

Without doubt the concentration of various climate-affecting trace gases in the atmosphere has risen significantly, which can only be attributed to the increased release of these gases by human activities. This is especially true of carbon dioxide, which has so far been responsible for about 50% of the artificial greenhouse effect. The other half comes from the greenhouse gases methane, nitrous oxide, ozone, and the CFCs, the chlorofluorocarbons, which are also known as "ozone-killers".

With their very long lifetimes, all these gases have accumulated in the atmosphere, and have thus increased their concentrations appreciably, as Figure 2 shows. One important greenhouse gas has to be dealt with separately, namely water vapor. Without this gas the earth's surface would be 20 degrees colder, as 2/3 of the natural greenhouse-warming effect of 30 degrees comes from water vapor. This natural greenhouse effect, which significantly reduces the earth's longwave radiation by means of the atmosphere absorbing and re-emitting this radiation first made life on earth possible. It is still unclear whether human activities have already significantly changed the water vapor content of the atmosphere.

There is no doubt that the ozone content of the stratosphere has steadily decreased in the last few decades (see Fig. 3). Over the Antarctic, some layers have no ozone left whatsoever for certain periods of the year. The hole in the ozone layer is becoming ever bigger and deeper. Fortunately, however, the required temperatures of −80 °C mean that this dramatic ozone depletion is still restricted to the uninhabited Antarctic and to a period of 4 to 6 weeks in the spring. However, densely populated areas of the northern hemisphere are already showing signs of steady ozone depletion in the stratosphere (see Fig. 4), although at 10% it has not yet reached extreme levels. At the same time, a steady increase in low-altitude ozone concentration in all industrial areas has been observed, as ozone smog, which occurs in urban areas during periods of intense sunshine.

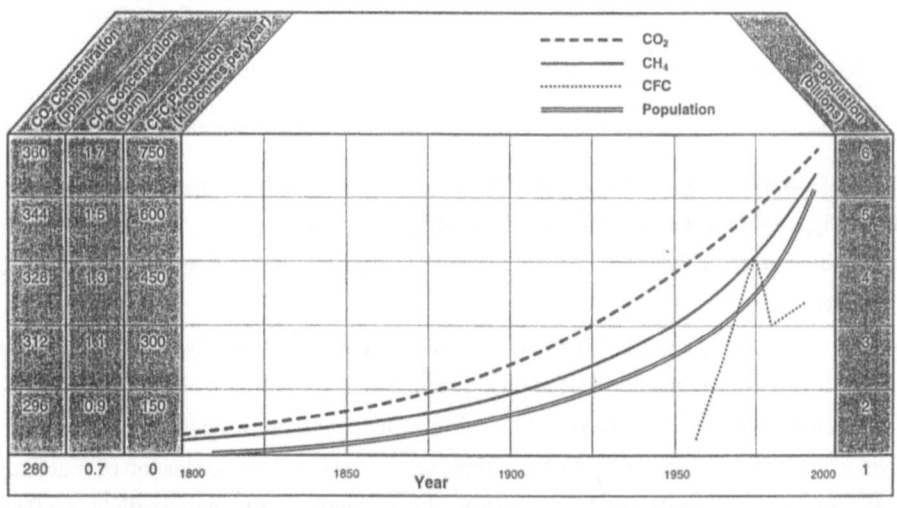

(Source: Enquête Commission German Bundestag, 1991)   REF/Geo 08/94

Figure 2. Increases in greenhouse gases and world population growth.

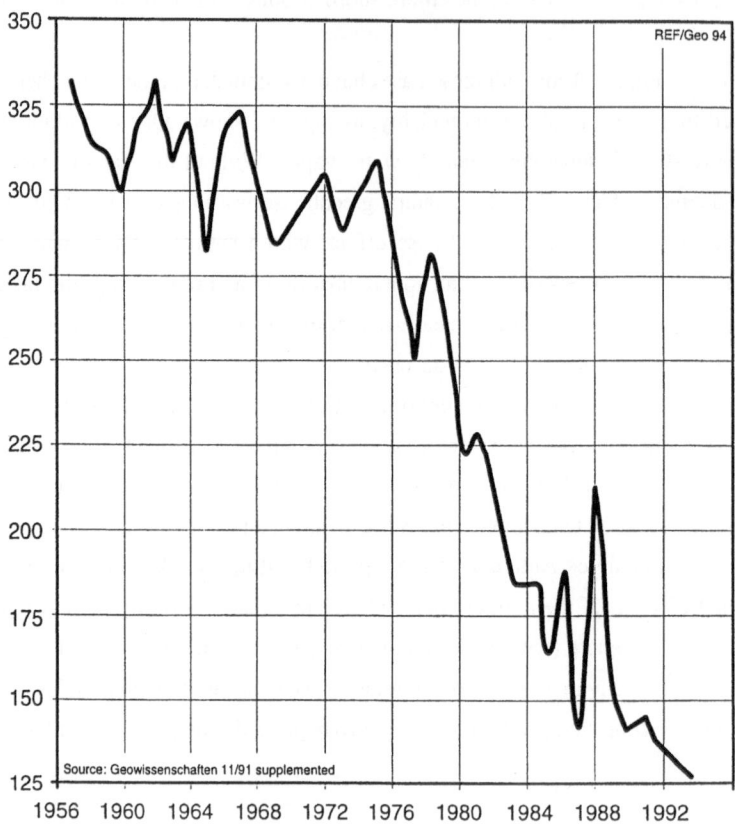

Figure 3. Ozone depletion over the Antarctic. (October means in Dobson units = 0.01 mm.)

While CFCs have been proved to be the main cause of the hole in the ozone layer, ozone smog is mainly triggered by car exhaust fumes. Both phenomena are positively interlinked, i.e. they intensify each other, so that an increase in the harmful effects on humans, animals and plants can be expected.

The air is becoming increasingly turbid throughout the world, especially in industrial regions. This is due to aerosols, tiny particles of dust, soot and sand, as well as droplets from condensed exhaust gases such as sulfur dioxide, as they diffuse sunlight more strongly and give the atmosphere a "milkier" appearance. At the same time, aerosols assist tile condensation of water vapor and the formation of clouds, which then appear whiter. Consequently, more sunlight is reflected back to space from the surface of the clouds and less reaches the earth's surface (see Fig. 5). This weakens the greenhouse effect during the day, or rather "disguises" its true strength. At

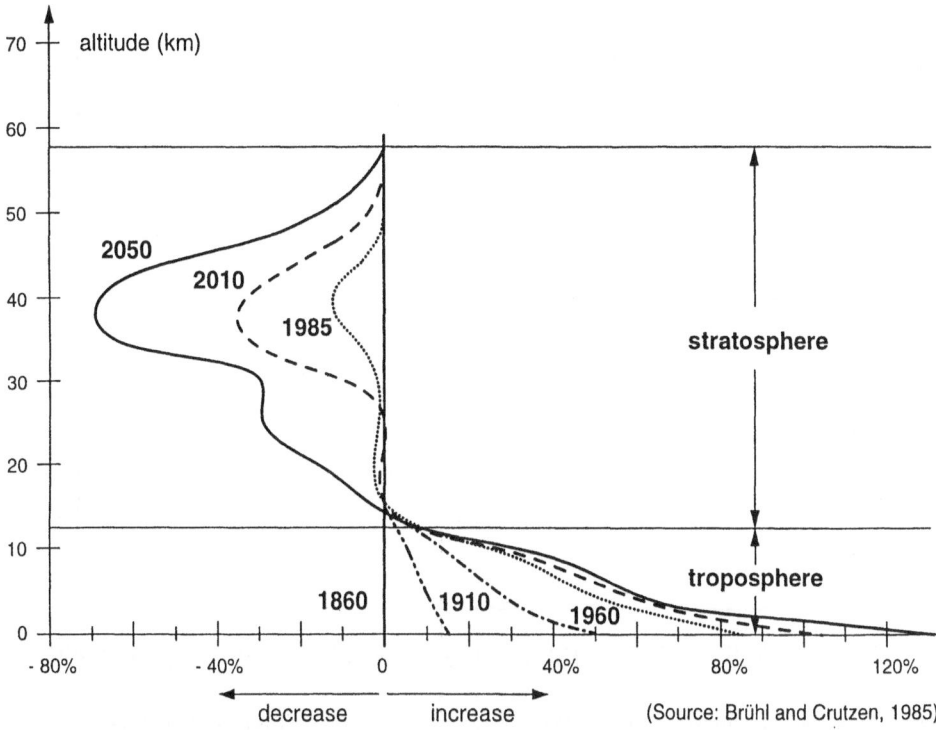

Figure 4. Atmospheric ozone content. Observed and predicted relative changes.

night, on the other hand, the longwave emission of the atmosphere to the earth is intensified by the aerosols, which reduces the cooling effect of the night. And in fact, an increase in night temperatures has been observed worldwide while daytime temperatures have only shown slight changes. It is a similar case with the course of annual temperatures: warming has so far mainly taken place in winter.

Natural phenomena, such as eruptions of volcanoes, can also temporarily reduce the radiative budget of the lower atmosphere. This was the case, for example, after the eruption of Pinatubo in 1991 in the Philippines when some 7 cubic km of ash and gases were catapulted to a height of more than 20 km and the aerosols in the stratosphere increased by a factor of 100 over large areas. This significantly reduced the shortwave radiation for several years and even the average global temperature temporarily sank by several tenths of a degree. The aerosols have since fallen back to earth or have been washed out by rainfall, so that temperatures will soon reach and exceed their previous level.

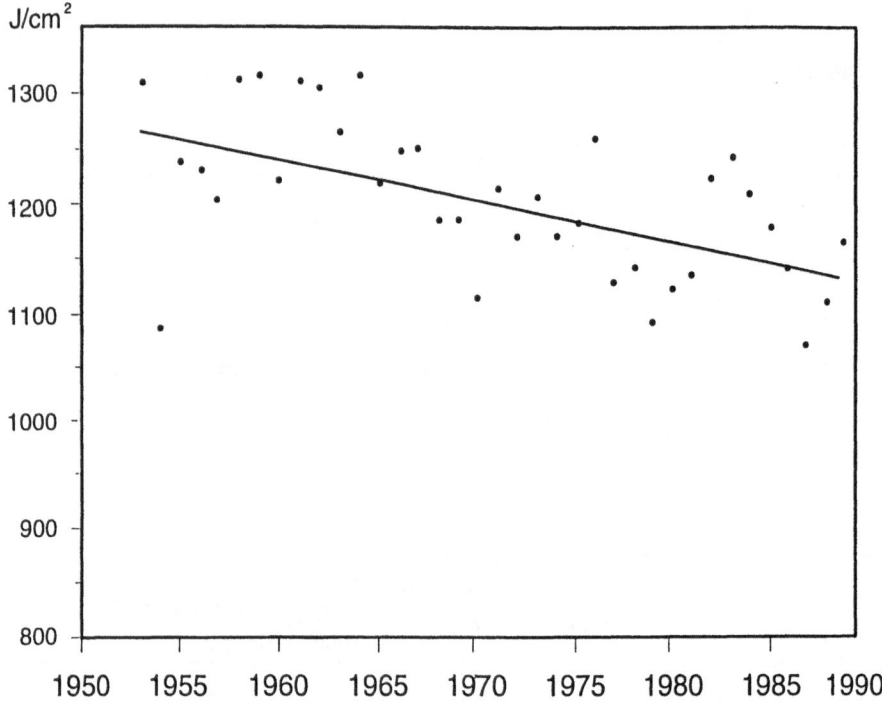

Figure 5. Mean annual shortwave radiation. (Meteorological Observatory, Hohenpeißenberg, Germany)

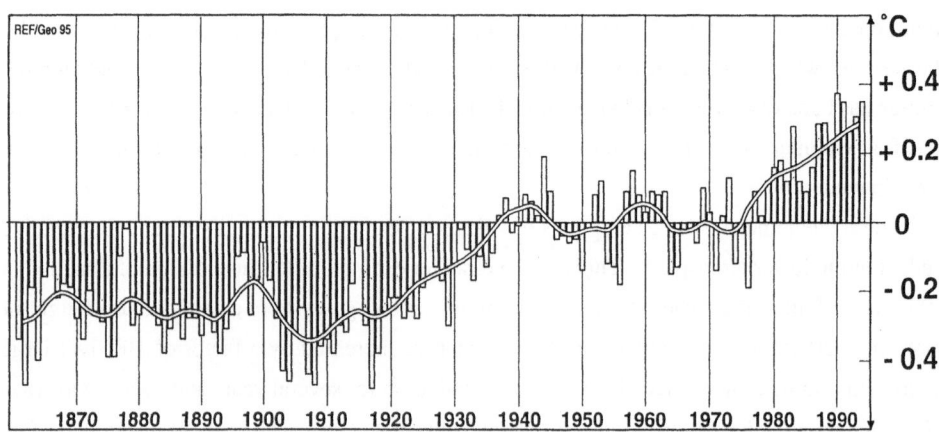

Figure 6. Global mean temperatures since 1861. Yearly deviations from average value (1951–1980) and running mean.

Severe melting of mountain glaciers has been observed worldwide, and has already assumed dramatic proportions in certain regions. Since their highest levels in the middle of last century, the Alpine glaciers have lost around 1/3 of their surface and half of their mass. The discovery of "Ötzi" is an obvious indication that the glaciers have not retreated as much for thousands of years. At the current melting rate, by the middle of next century there will only be lamentable traces of the once great Alpine glaciers. Large parts of Europe will thus lose their most important source of drinking water.

The melting of the glaciers causes the sea level to rise. Added to this is the thermal expansion of the sea water due to rising temperatures. The total rise in the sea level this century is 10 cm, an amount that may seem small but which does give cause for concern when it is seen as an indication of what is to come in the future.

The global rise in temperature of approximately 0.5 degrees (see Fig. 6) also appears harmless at first even if the last decade produced the highest temperatures since worldwide meteorological records began 130 years ago. It is only when one considers that the differences between the cold and the warm periods in recent climatic history were on average only 5 degrees that the real significance of this warming becomes apparent. Added to this is the fact that the temperature rise in the lower atmosphere contrasts with an even sharper temperature drop in the stratosphere (the energy in the lower atmosphere is absent in the stratosphere), which reduces the stability of the stratification of the atmosphere and at the same time increases the destruction of the ozone layer.

There is a considerable delay before ocean temperatures follow the temperature rise in the atmosphere. This is because the oceans can store a great deal of warmth, and deep-sea water circulates on a type of huge conveyor belt roughly only once every 10,000 years. Nevertheless, significant rises in temperature have been observed even in large ocean areas – most recently in the western Pacific warm water pool, which acts as a gigantic heat source. Higher ocean temperatures lead to exponentially higher evaporation and to a correspondingly higher water vapor content in the atmosphere. This assists the development of high rainfall intensities and tropical cyclones.

Warming is in no way evenly spread across the earth, indeed there are numerous regions which have even cooled down. Consequently, in some areas, such as over the northern Atlantic, the spatial differences in temperatures and air pressures have grown (see Fig. 7), resulting in a significant increase in windstorms in these areas in the last few decades. At the same time, winters in Europe have become milder with the result that the eastern European high-pressure system, which builds up in winter over the large snow-covered surfaces of Europe, has weakened on average and retreated eastwards (see Fig. 8). It is thus losing more and more of its blocking effect on the low-pressure systems approaching from the Atlantic, which are now able to penetrate more frequently and further eastwards and hit Europe with their hurricane force winds, as was the case

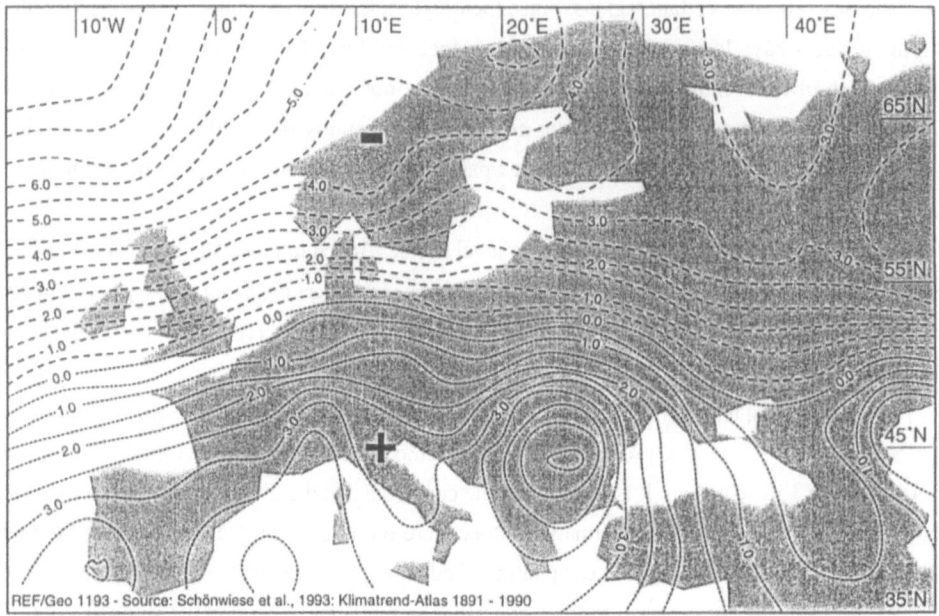

Figure 7. Trend of atmospheric surface pressure (hPa); winter (1961–1990).

in January/February 1990. This is of greatest concern to the insurance industry since an increased frequency of similar conditions could virtually ruin some sectors of the affected markets.

Many other changes have affected the earth, such as the expansion of deserts and the increase in droughts on the edge of these areas, remarkable changes in flora and fauna – from increased wintering of former migratory birds in our latitudes, to dying forests – and not least the increase in tropical diseases outside the areas where they are normally found. Climate change is usually only one of the many reasons for the numerous changes that are taking place in environmental and living conditions.

The decisive question is not whether this long list of evidence is conclusive but whether the climate data and computer climate models can provide enough information to assess future changes and develop the appropriate adjustment and preventive strategies in time. The risk of error will remain great for the foreseeable future. It is therefore all the more important that the strategies themselves can be adjusted and do not lead down any blind alleys. So-called "no-regret" strategies, such as the reduction of fuel consumption for cars or the reduction of energy consumption in general, are successful from the very outset as (even if the relevance to climate is

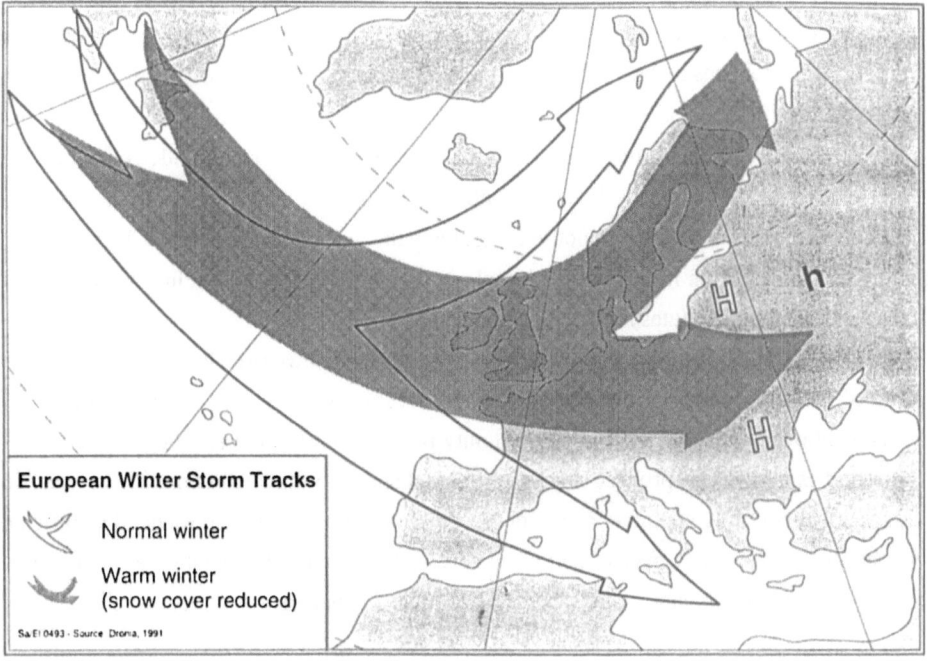

Figure 8. European winter storm tracks.

lower than expected) they result in desirable savings and are also very useful to demonstrate to the Third World the industrialized countries' sense of responsibility.

4. Climate change predictions

What sort of future do climatologists predict? The most reliable forecasts seem to be those made by the Intergovernmental Panel on Climate Change (IPCC), as the many climatologists from all over the world that work together in this group are much less likely to make fundamental errors in their statements than individual scientists, who often have quite contradictory opinions. These conflicting opinions should not in any way be dismissed, as they often quite rightly take issue with factors that have a considerable degree of uncertainty. Unfortunately, however, they also serve to provide decision makers in the political and economic worlds with convenient and at the moment most welcome excuses for a failure to act or a failure of policies.

The first IPCC report (IPCC, 1990) and its supplement (IPCC, 1992) extrapolate future development on the basis of various scenarios, of which the most plausible, "the business as usual" scenario, is considered the worst possible case. But is this really true? In view of the rising economic problems in the most heavily populated regions of the Third World, should it not be expected that the release of greenhouse gases will rise even more quickly in the future than has so far been the case? If this is true, the expected doubling of the carbon dioxide content by the middle of next century could well be too optimistic. There is also serious doubt about the assumed growth rates of other greenhouse gases. Only the agreed phasing out of CFCs seems to be going as planned, even if new production plants are going into operation in some countries, and the safety of the substitutes is not yet guaranteed.

According to the IPCC, the average global temperature will have risen by several degrees by the end of the next century. However, the lower curve in Figure 9 does seem unrealistic for reasons mentioned above. The area of uncertainty ranges from about 2 to 5 degrees. The next IPCC report is unlikely to produce any substantial changes to this prediction. This means an average global rise in temperature of 0.3 degrees per decade and thus represents a significant acceleration over the 0.5 degrees in 100 years observed so far. The acceleration is so great that many ecosystems will probably not be able to adjust quickly enough.

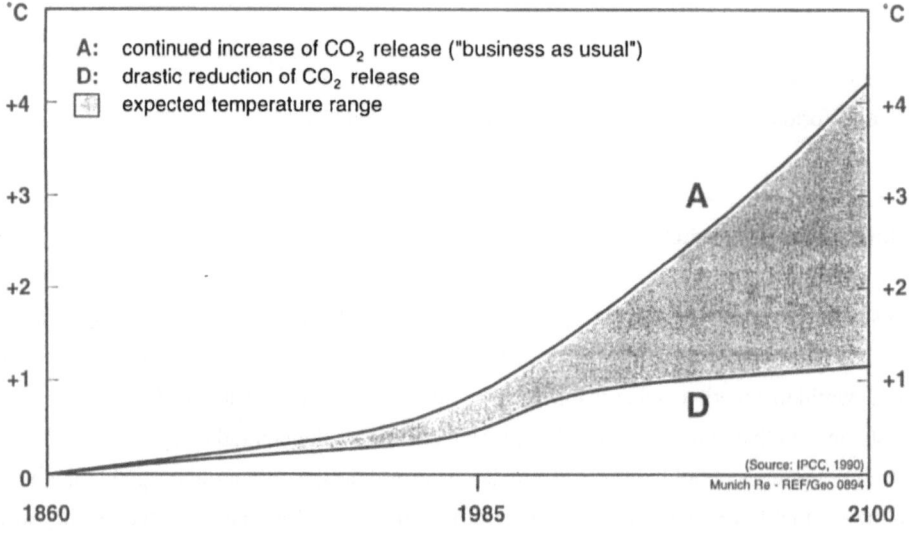

Figure 9. Global rise in surface temperature.

The above mentioned thermal inertia of the oceans and the buffer effect of the increase in snowfall in the Antarctic will cause the sea level to rise slightly more slowly than the previously assumed level of approximately 65 ± 30 cm in 100 years. This will not change the dramatic melting of inland glaciers in most mountainous regions of the world. Equally dramatic is the increase in humidity as a result of increased evaporation, as this has a decisive influence on precipitation and convective processes. We can therefore not only expect more torrential rain, flash floods and mudflows but also more thunderstorms, hailstorms, lightning and tornadoes. Tropical cyclones – including hurricanes and typhoons – are not only likely to increase in intensity but are also likely to extend their seasons and areas of occurrence considerably. Extratropical storms, i.e. winterstorms, are also likely to become more severe and, as explained above, penetrate further inland. In conjunction with the rise in sea level, increased windstorm activity will result in a sharp increase in the risk of storm surges for many coastal areas (Berz and Conrad, 1994). Table 3 summarizes all these effects.

Table 3. Climatic risks for the insurance industry

Increase in

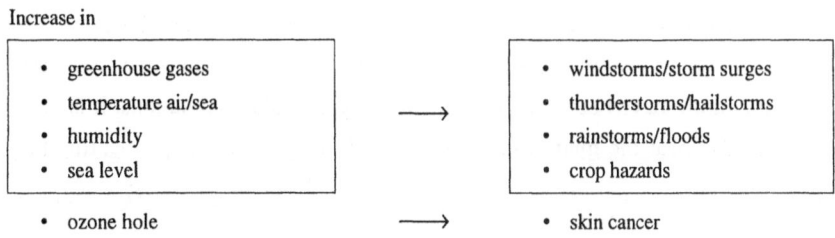

• greenhouse gases	• windstorms/storm surges
• temperature air/sea	• thunderstorms/hailstorms
• humidity	• rainstorms/floods
• sea level	• crop hazards
• ozone hole	• skin cancer

5. Economic effects of climate change

As climatic history can show us, a warmer global climate generally means more precipitation. In the long-term the moderate and subpolar regions will profit from this while the areas which form the breadbaskets of the world will suffer from more frequent droughts. Even if the rising CO_2 content of the atmosphere improves growth conditions for most plants and their water requirements fall, this will be outweighed in many countries by the negative effects on agriculture, chiefly due to the exacerbated effect of heat and aridity. The winners will be those countries that can adapt their agriculture most quickly to the changing conditions; these are obviously the highly developed countries, while the less developed ones will fall further and further behind.

However, not all effects have to be negative. For example, many countries in the moderate and subpolar regions of the world will be able to reduce their heating costs radically, and the risk of frost will also be greatly diminished. Against this, other regions will see summertime energy consumption for air conditioning systems rise significantly and more frequent heatwaves will cause additional damage.

Attempts have been made recently to estimate the worldwide costs of the artificial greenhouse effect and to take appropriate stabilizing measures. The results are quite clear. According to investigations by the Fraunhofer Institute for Systems Engineering (Hohmeyer and Gaertner, 1992) the probable losses from climate change will increase up to a figure of some 1.5–5% of global GNP per year by the end of the next century. However, permanent preventive strategies would reduce this by half. In spite of this, many governments and economic enterprises still seem to be undecided about or even opposed to the Framework Convention on Climate Change signed in Rio in 1992. The further negotiation process will show whether the 166 signatory states are really serious about their decision regarding measures to protect the climate.

An interesting question could be which climate should be sought as the most favorable for the whole earth. On the other hand, there is absolutely no question that the global experiment that has been attempted in such an aimless and uncontrolled fashion will have to be brought under control as soon as possible if later generations are not to inherit a completely destabilized climate. If the problem of climate change is seen from this ethical point of view, there can be no doubt whatsoever that it is high time something was done. As the climate changes witnessed so far and those predicted for the future have been caused in the main by the industrial nations, it is only right that they should also bear the main responsibility for developing and implementing the necessary countermeasures.

6. Conclusions for the insurance industry

Confident in its exceptional ability to adjust to changing risk circumstances, the insurance industry could now adopt the attitude that it is not really affected by climate change. However, I must warn strongly against such an attitude. Climate change will produce, in nearly all regions of the world, new extreme values of many insurance-relevant parameters that will lead to natural disasters of unprecedented severity and frequency. This will cause capacity problems in national and international insurance markets that are much more serious than those experienced in the last few years. The whole future of this class of business in certain regions could be at stake if the development of this problem is misjudged. Premium income would also lag behind loss development in this case. The climatic changes have significant impacts on the insurance industry:

- increase in weather variability,
- new extreme values in certain regions,

- new exposures,
- more frequent and larger natural disasters,
- greater claims potential,
- poorer claims experience,
- lagging premium adjustment.

On the other hand, it is possible for the insurance industry to protect itself adequately against the effects of climate change and, at the same time, make a significant contribution to implementing measures to protect the climate. No other sector of the economy has such effective instruments for encouraging risk reduction. However, this can only work if the insurance industry can have its clients and the authorities as useful partners on its side. If, for example, insurers can convince their clients that substantial deductibles in natural hazards insurance are of benefit to both sides because they relieve insurers from the great number of minor losses, which can be more effectively remedied by clients, and because they substantially reduce the price of coverage, clients will then be more willing to take measures that prevent or minimize losses. If, on the other hand, the insurance industry looks at the growing loss potential and decades to exclude certain perils or risk areas from cover or severely restrict the scope of cover, then the pressure will inevitably grow on authorities to take measures to improve risks or to tackle the cause of losses or even to transfer the risk from the state welfare system to the government. The proper response of the insurance industry to the deteriorating environmental conditions must include:

- sound technical pricing,
- significant deductibles,
- appropriate liability limits,
- transparency / accumulation control,
- loss prevention / mitigation,
- awareness / education.

It would, however, be wrong if the insurance industry were to take on the state's role of penalizing or rewarding environmentally-damaging or environmentally-friendly conduct on the part of its clients. This cannot be the task of the insurance industry even if it were in its interests with regard to the correlation between environmental losses and natural disasters.

Humankind is carrying out a huge experiment with the earth's climate, an experiment over which it has so far had practically no control and the outcome of which is still very much open. It could, however, dramatically affect human living conditions in the future. As much as people can still argue today about the future development of climate change and in particular the effects it will have, this very high level of uncertainty is itself a great cause for concern. Even if the extreme caution we are now showing proves to be unfounded, we have no choice if we are to behave responsibly towards later generations. We must try with all available means to ensure that, meta-phorically speaking, the greenhouse window does not close even further.

References

Berz, G (1993) Global warming and the insurance industry. *Interdisciplinary Science Reviews* 18(2):120–125.
IPCC (Intergovernmental Panel on Climate Change) (1990) *Climate Change: The IPCC Scientific Assessment.* Cambridge University Press, Cambridge.
IPCC (Intergovernmental Panel on Climate Change) (1992) *Climate Change: The Supplementary Report to the IPCC Scientific Assessment.* Cambridge University Press, Cambridge.
Berz, G and Conrad, K (1994) Stormy weather: The mounting windstorm risk and consequences for the insurance industry. *Ecodecision* 12:65–68.
Hohmeyer, O and Gaertner, M (1992) *The Costs of Climate Change.* Report to the Commission of the EU. Fraunhofer Institut für Systemtechnik und Innovations-Forschung, Karlsruhe.

Part II

Stabilization targets, costs, and technologies

European stabilization targets: What do they bring, how much do they cost?

Eberhard Jochem

Fraunhofer Institute for Systems and Innovation Research, Breslauer Str. 48, D-76139 Karlsruhe, Germany

This paper is a synthesis of several observations, not many of which have been published or discussed in detail, covering differences in climate change policy between North America and Western Europe (of small and rich European countries in particular), a systematic overestimation of mitigation cost by additional energy efficiency improvements in energy models, and a cost evaluation of climate change policy in Germany where some 5% of the gross domestic product (GDP) increase would be needed to meet the 25% emissions reduction target by 2005, a small price in the eyes of the author for reducing the anticipated risks.

1. European stabilization targets: The leading climate change policy

There is sufficient information on the quantitative targets of the European Union (EU) and its member countries (Jochem and Henningsen, 1994). The CO_2 emissions of EU member countries are likely to increase by some 10% between 1990 and 2000, taking into account a decrease of 20% in the former GDR (see Tab. 1). The majority of the EU member governments have set some emission targets for the year 2000 or 2005, and even countries with lower GDP *per capita* expect to be able to limit their growth of CO_2 emissions (Ireland, Greece, Portugal, Spain, see Tab. 2). Although the EU stabilization target of approx. 9 tons of CO_2 *per capita* seems to be achievable, some doubts remain that this target will be met. The doubts shared by many analysts are mainly due to an ever increasing mobility on European roads, the lack of car efficiency and transport policy in the European member states, and the hesitation to introduce an energy/CO_2 tax at the European level, on the grounds of a loss of competitiveness of the European energy-intensive industrial branches.

Nevertheless, besides the quantitative stabilization target, the climate change policy of a few Western European countries is rather active and responsive to possible risks. Whereas the North American governments argue that cumulative CO_2 emissions have to be considered, which will not change very much in the next decades independent of whatever rigid mitigation policy will be realized, the position of many Western European governments is that the increase in CO_2 concen-

Table 1. CO_2 emissions: 1990 levels, national emission targets and CEC reference projections from 1992 analyses

EC member country	Emissions in 1990			Emissions in 2000	
	Total (million tons)	Tons per capita	% EC	National targets (million tons)[a]	CEC projected (million tons)
Belgium	112	11.2	3.7	106	122
Denmark	53.1	10.3	1.7	50	65.5
France	366	6.5	12.0	425	431
Germany	1008	12.6	33.1	<885	1037
West	708	11.1	23.3	610	800
East	300	18.8	9.8	<275	237
Greece	73.7	7.3	2.4	92	96.6
Italy	402	7.0	13.2	400	464
Ireland	30.8	8.8	1.0	37	36
Luxembourg	12.5	32.7	0.4	13	13.7
Netherlands	157[b]	10.5	5.2	150[b]	178
Portugal	39.9	4.1	1.3	55	57
Spain	211	5.4	6.9	263	260
United Kingdom	579	10.1	19.0	587	614
EU 12[c]	3045	8.9	100	3063	3375

[a] Figures for countries that have a year 2005 target estimated.
[b] Contradiction between Dutch and EC sources.
[c] Total figures are rounded and include East Germany.
(Jochem and Henningsen, 1994).

tration in the atmosphere per decade or per century will be decisive for harmful or acceptable impacts on vegetation, migration and national economies. Therefore, many Western European governments at the national, regional or city level are implementing policies of CO_2 emission mitigation (e.g. Denmark, the Netherlands, or regions like Flanders in Belgium, North Rhine-Westphalia, Schleswig-Holstein, Hessen in Germany). Whether this difference in policy of the early 1990s is due to scientific disagreements or the unwillingness to commit themselves to some targets in North America is an open question. Policy positions and the selective acceptance of scientific results may interact in both cases in the "old and new world". In Europe the arguments of no-regret investments and additional impact on innovation seem to be more accepted than in North America. As the reduction of greenhouse gas emissions depends very much on improvements in energy efficiency, the relatively active policy in many Western European countries is likely to strengthen the competitiveness of their economies in producing energy-intensive goods in the longer-term, and in foreign trade of energy-efficient investment goods (Hohmeyer and Jochem, 1992).

Table 2. CO_2 emission targets of national governments and of the European Union and adopted target values for 2000

EC member country	Base year	Target year	Target (%)	Adopted target 2000 (million tons)
Belgium	1990	2000	0	106
Denmark	1988	2005	−20	50
France	—	—	0[b]	425
Germany	1987	2005	−25 to −30	<885
West	1987	2005	−25	610
East	1987	2005	−25 to −30	<275
Greece	1990	2000	+25	92
Italy	1990	2000	0	400
Ireland	1990	2000	+20	37
Luxembourg	1990	2000	0	13
Netherlands	1989	2000	−3 to −5	173–182
Portugal	1990	2000	+29 to +39	55
Spain	1990	2000	+25	263
United Kingdom	1990	2000	0	587
EU 12[a]	1990	2000	0	~3100

[a] Total figures are rounded and include East Germany.
[b] *Per capita* stabilization of 7.3 tons.
(Jochem and Henningsen, 1994).

2. Lack of knowledge about the realistic costs of reducing CO_2 emissions

Comparative analyses of the average additional investment cost for additional energy efficiency improvements or substitution, disaggregated for individual branches and fuels, show high deviations, depending on whether they stem from empirical engineering analyses or cost estimation methods (see Tab. 3 and Jochem and Bradke, 1996). The major criticisms made by the author of the commonly used energy-saving cost curves are (see Fig. 1):

- The potentials often include no information about the market imperfections and obstacles which might hinder the full economic efficiency potential from being realized and which may be substantial in many sectors (Gillissen et al., 1995).
- Sometimes, the efficiency (or substitution) potentials are limited by the authors to a given cost level and implicitly suggest an increasing marginal cost to the nontechnologist, which is not the case in many instances, due to complete process substitutions, car substitution concepts or different construction techniques (e.g. Jochem and Bradke, 1996).

Table 3. Additional savings and investments in the tertiary sector compared with the reference scenario

Literature	Reduced energy[1] demand compared to reference [PJ]	Cumulative additional investments [Billion DM]	Average specific additional investments per year [Billion DM/PJ]
Altner et al. (1995)			
Target-I Alternative	160	43.9	0.274
Target-II Alternative	235	72.3	0.308
Schön et al. (1994)			
Scenario R1, 2020	247	70	0.283
Scenario R2, 2020	265	84	0.317
Jochem (1994)			
2010	160	45	0.283
IER (1995)			
Target-I Alternative	156	57.5	0.368
Target-II Alternative	231	120.8	0.523

[a] Total Final Energy
(Nitsch, 1995)

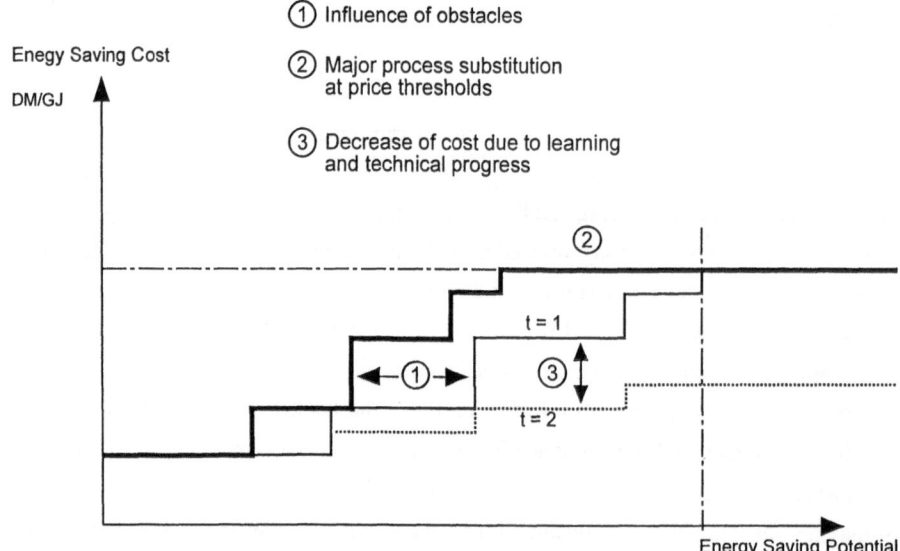

Figure 1. Limitations of energy saving / cost curves, particularly in the industrial and tertiary sectors (Jochem and Bradke, 1996).

- Learning curves and autonomous technological progress reduce the specific mitigation cost in time; this reduction may be substantial in areas of newly applied technologies.

These effects on the CO_2 mitigation cost are often not considered thoroughly enough by bottom-up optimization models and hardly considered at all by top-down models. The experience of the author is that because of these unknown data specifications, the figures on cost and efficiency potentials are taken for granted and the cost figures are very often far too high, future cost figures in particular. Unfortunately, there is no literature which takes up this issue, including the aspects of transaction costs and program costs, which may not be additive in many cases (e.g. in contracting energy converting plants by specialized energy service companies from other companies or public bodies).

In addition, intrasectoral structural changes, which have gained importance in some sectors, such as the basic product industries, in many European countries during the last ten years, or which will become more important in the next few decades, and saturation effects in energy-intensive consumer sectors, cannot be adequately handled in top-down models.

3. What are the additional costs of CO_2 mitigation in the next 10 to 50 years?

Calculations have been made using an optimization model (Bradke et al., 1991; Tönsing et al., 1996) for which the technical and cost data were collected and estimated in close cooperation with German manufacturers, engineering companies and applied research institutes. The model is a linear optimization model with some 140 technical areas to simulate the energy use, emissions and related costs in the final energy and energy conversion sectors. The influence of data uncertainties was considered by varying technical and cost data by a systematic lower and upper end of two scenarios. The higher the target on CO_2 emissions, the less important were the uncertainties of longterm cost estimates for the technical structure of the energy system. With higher CO_2 reduction targets set for the middle of the next century, the cost of the German energy system will increase, of course, but even in the -75% reduction case (based on 1990), the cost share of the energy system (including all additional costs in the final energy sectors) is reduced from 8.2% of GDP to around 6% by 2050, including a doubling of the cost element of energy imports and energy transmission and distribution (see Fig. 2). At first glance, these additional energy systems costs may seem rather low. But the figures reflect several effects of a highly industrialized country: autonomous (and additional) energy efficiency improvements, structural changes to less energy-intensive products and services, saturation of material-intensive consumption, increasing recycling and re-use of products and wastes, saturation in mobility by 2050.

Rough estimates have been made for Germany and the target year 2005 derived from these data and optimization calculations (see Tab. 4). Assuming an annual average *per capita increase*

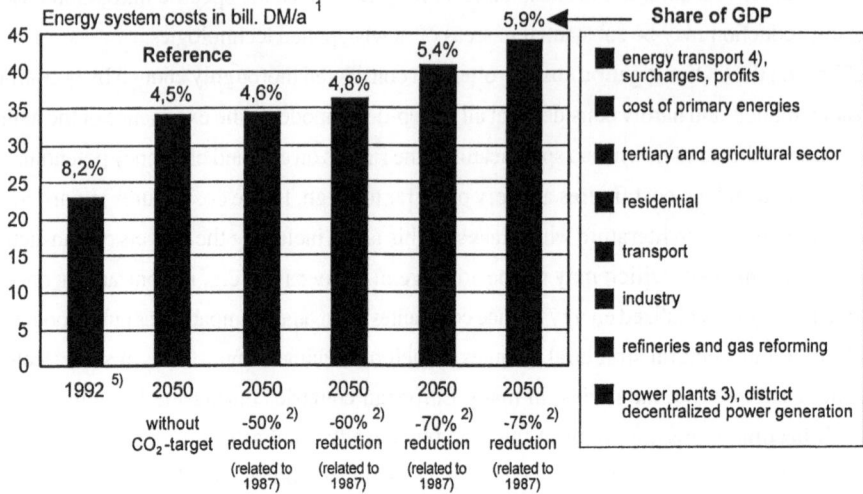

Figure 2. Energy system cost of Germany in 1992 and 2050 with different reduction targets in CO_2 emissions.
1) Capital cost plus O&M cost of power plants, boilers; differential cost in the final energy sectors between target and reference case.
2) Including industrial power plants.
3) Photovoltaic plants, fuel cells in the residential and tertiary sector.
4) Includes grids for electricity, district heat, gas, hydrogen and oil.
5) 1992: investment cost of electricity and district heat generation, cost for primary energy production and imports as well as profits are estimated.

Table 4. GDP and energy system cost of Germany in 1992 and 2005, Reference and Policy case, *per capita* figures

Year	Variable	Total $(10^{12}$ DM$)^a$	*Per capita* (DM/cap)	Share of GDP
1992	GDP	2.9	36,000.-	100 %
	energy system cost	0.24	2,960.-	8.2 %
2005	*Reference case*			
	GDP (2.2 %/a)	3.80	46,900.-	100 %
	• increase 1992/2005	0.90	840.-	
	energy system cost	0.25	3,090.-	6.5 %
	• increase 1992/2005	0.01	10.-	
2005	*Climate Policy case*			
	energy system cost	0.295	3,640.-	7.8 %
	• increase 1992/2005	0.055	52.-	

^{a} at prices of 1991

of GDP of about 840.— DM per year and per capita (a.cap) (which has also been observed during the last few decades in West Germany), some 10.— DM/a.cap of this increase would be necessary to finance the reference case (increased shares of gas and electricity) and about 52.— DM/a.cap in the climate policy case, which would allow the German government's CO_2 emission reduction target of −25% by the year 2005 to be met.

The decision which has to be made is whether to spend 6% or 1% of the additional GDP on a greater or lesser climate change in the next decades. Whether these additional costs are considered to be rather low or unbearably high by the citizens in the OECD countries varies according to the different cultures and interests involved. Looking at the Western European countries, there is some evidence that the citizens of the small and rich countries (such as Denmark, the Netherlands, Sweden and Switzerland) are prepared to pay the extra cost for contributing to changing today's economic paradigm which concentrates on short-term benefits, and to contribute to "limits to climate change". These small European countries perform at good levels of competitiveness in the know-how intensive capital goods industries and fine chemicals. They are not afraid of the "free riders" among the OECD countries which have not yet set targets; they focus on their chances, on the long-term trends and demands of a future cyclic economy with its vast changes in capital stock, technology and behavior during the next century. They consider their strategy to be an economic opportunity and an obligation for future generations.

4. Final suggestions for research

Countries and different societal, scientific and political systems learn from each other. There are common concerns about more understanding and more precise information, which could be improved by *empirical research*:

- The future cost of energy-efficient technologies which will not be "add on" technologies, but new construction techniques, new industrial processes (University of Utrecht, 1994) and new vehicles and transport systems. What are the shapes of their learning curves?
- What are the valid reasons for the fears of energy-intensive industries about loss of competitiveness? How could they be considered in energy/CO_2 surcharge schemes? Why are the winners, the capital goods industries and the biotechnology and membrane suppliers, so silent today?
- What are the efficient instruments (or bundles of measures) of local and national governments and promising company strategies (e.g. contracting/outsourcing) to overcome existing obstacles to no-regret investments to reduce CO_2 emissions in various target groups?

The usefulness of existing models (whether designed top-down or bottom-up) and policy arguments is limited because they simplify reality. The author therefore suggests increasing activity in the above mentioned fields of empirical research.

References

Bradke, H, Brakhage, A, Knöpfel, I, Lyons, B, Reichert, J, Treber, M, Masuhr, KP and Eckerle, K (1991) Konsistenzprüfung einer denkbaren zukünftigen Wasserstoffwirtschaft. Untersuchung im Auftrag des Bundesministeriums für Forschung und Technologie, Bonn.
Gillissen, M, Opschoor, H, Farla, J and Block, K (1995) *Energy Conservation and Investment Behavior of Firms.* Dept. Science, Technology and Society, University of Utrecht, Utrecht.
Grubb, M., Edmonds, J., ten Brink, P. and Morrison, M. (1993) The costs of limiting fossil fuel CO_2 emissions: a survey and analysis. *Ann. Rev. Energ. Environ.* 18:397–478.
Hohmeyer, O and Jochem, E (1992) The economies of near-term reductions in greenhouse gases. *In*: IM Mintzer (ed.): *Confronting Climate Change.* Cambridge University Press, Cambridge, 217–236.
IPCC (Intergovernmental Panel on Climate Change) (1995) Summary for Policymakers: *Economic and Social Dimensions of Climate Change.* Montreal, Canada.
Jochem, E and Bradke, H (1996) Energieeffizienz, Strukturwandel und Produktionsentwicklungen in der deutschen Industrie. IKARUS-Monographien, Forschungszentrum Jülich.
Jochem, E and Henningsen, J (1994) How Europe convinced itself that it could meet the carbon dioxide stabilization target. *Ann. Rev. Energ. Environ.* 19:365–385.
Nitsch, J (1995) *Anmerkungen zur Studie 'Zukünftige Energiepolitik – Vorrangig für rationelle Energienutzung und regenerative Energiequellen.* DLR, Stuttgart.
Tönsing, E, Jochem, E, Landwehr, M, Mannsbart, W, Nölscher, C, and Vollmar, H (1997) Szenarien der Energienutzung und -versorgung mit reduzierten CO_2-Emissionen bis 2050. *Energiewirtschaftliche Tagesfragen*: Forthcoming.
University of Utrecht (1994) Long-term industrial energy efficiency improvement: Technology Descriptions. NOVEM, Sittard, The Netherlands

Cost-Benefit Analyses of Climate Change: The Broader Perspective
F.L. Toth (ed.)
© 1998 Birkhäuser Verlag Basel/Switzerland

Climate protection and the economy of prevention

Peter Hennicke

Wuppertal Institute for Climate, Environment, Energy GmbH, Döppersberg 19, D-42103 Wuppertal, Germany

The following discussion debates the issues involved in the longer-term analysis of the energy system in the stricter sense (i.e. excluding transport and nonenergetic energy consumption), and the possibilities of reducing energy-related CO_2 emissions. Three pivotal questions shall be addressed. (1) Which technological options are available to climate protection policy? (2) Is the further propagation of risks, e.g. the acceptance of more nuclear risks, inevitable if the world's climate is to be protected? (3) Is it possible to finance a risk-minimizing energy strategy that avoids both the risks inherent in harnessing nuclear power and those posed by climatic changes?

On the basis of reported scenarios, these questions shall be answered at the geographical levels of the world, the European Union and the Federal Republic of Germany. The analysis reveals that climate protection policy is broadly equivalent to proactive industrial policy. In other words, even if the urgency of comprehensive climate protection measures were not accepted by the political and business communities, both macroeconomic arguments and the precautionary principle provide reason enough to pursue an active climate protection policy. The extra costs of an aggressive climate protection policy are far lower than the climate-related damages potentially entailed by a business-as-usual energy policy. The prime strategy must be to apply an appropriate mix of instruments to reverse the perverse incentive structure that presently prevails in the energy system. It is not the additional consumption of energy and rising pollutant emissions which should, as is today the case, be worthwhile for suppliers and consumers, but rather the provision of additional energy services with less consumption of (nonrenewable) energy. We term such a system "the economy of prevention".

1. Global scenarios: Wide range of conceivable 'energy futures'

Energy scenarios are simplified, model-type representations, constructed as consistently as possible, of technically feasible future energy systems. They depict and make clearer the interactions in complex systems, and illustrate – by means of varying the assumptions made and objectives set – strategic alternatives and the available leeway for action. By contrast, forecasts attempt to make statements about the expected actual development. Scenarios can serve as a basis for forecasts, but do not as a rule claim to anticipate real developments. The lack of accuracy demonstrated in the past by energy forecasts is particularly indicative of the necessity of making a clear distinction between scenarios (conceivable alternative future pathways) and forecasts (prediction of likely development).

The history of 'official' energy forecasts is marked by a systematic and – at times – boundless overestimation of future energy consumption levels (see Müller and Hennicke, 1995). This has had two principal causes. *First*, energy forecasts 'at the upper edge' can be used by corporate boards as vehicles to convince the state, shareholders and regulatory authorities of the necessity of granting or authorizing public subsidies or investment budgets for new supply-side capacities and large-scale technologies (breeder, fusion). Energy futures are thus generated 'from the seller's perspective'. On the highly centralized and highly powered energy markets, the current framework conditions create a direct business incentive to plan at the upper edge of conceivable energy consumption. To date – and particularly in the electricity sector within regional supply monopolies – it has been possible to pass the costs of the overcapacities thus created on to the customers.

Second, energy experts for decades have taken it as an unwritten law that energy consumption grows by at least the same rate as the economy. Right into the 1970s, energy forecasts gave scant attention to the questions of which services energy was actually needed for, and how efficiently this energy was converted into such services. The oil companies flooded the OECD countries with cheap Near East oil, and the low real energy price level led to the establishment of energy-wasting structures in all industrialized countries. To the surprise of the experts, a 'decoupling' of economic growth and energy consumption occurred after the 'oil price crises' of the 1970s. Since then, it has been accepted that energy consumption can remain constant or even sink despite economic growth; nonetheless, the claim continues to be made that a longer-term decoupling is impossible (on the assumption that savings potentials will soon be exhausted), and that electricity consumption at least will continue to rise in step with economic growth.

Towards the end of the 1970s, the schools of thought in energy forecasting began to polarize sharply. Talk was made of 'hard' versus 'soft', or 'supply-oriented' versus 'use-oriented' energy paths. Greater attention began to be paid to the distinction between scenarios and forecasts. Amory Lovins' proposition of an 'efficiency revolution' was much vaunted, and his programmatic book, "Soft Energy Paths – Towards a Durable Peace" (1977) gave a worldwide impetus to a whole series of 'alternative', use-oriented energy scenarios. But only since the 1990s has the energy science establishment no longer called into question the fact that substantial technically realizable efficiency potentials remain to be tapped in all countries. Nonetheless, ruling energy policy continues to view forceful energy conservation as a technically unattractive, unreliable and above all unfinancable resource. But this – perhaps the most stubborn – preconception, too, is proving increasingly untenable.

1.1 Rising energy consumption is not a matter of fate, but political will

Since the early 1980s, the divergence that has developed between the various global 'energy futures' thought by experts to be technically accessible has been astonishing. Today we find at the level of scenario analysis an astounding 'peaceful coexistence' of 'hard, supply-oriented' and 'soft, use-oriented' strategies. Energy policy has as yet barely reacted to this divergence. This is all the more surprising as the unemotive matrices of facts and figures used in the scenarios lead to images both of frightful pathways into disaster, and risk-minimizing transitions towards sustainability.

A look at representative global energy scenarios published since the 1980s (see Schüssler and Hennicke, 1994 and Fig. 1) shows a range of technically accessible energy futures with

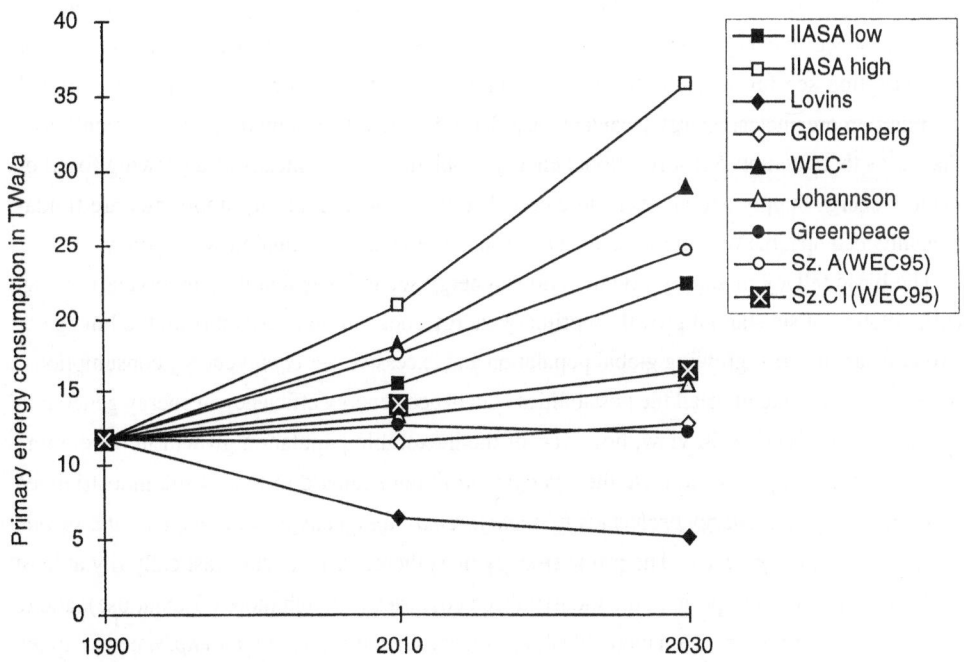

Figure 1. Overview of global energy scenarios (Häfele et al., 1981; Lovins et al., 1983; Goldemberg et al., 1988; Johannson et al., 1993; Greenpeace, 1993; WEC and IIASA, 1995; Grübler et al., 1995).

energy consumption levels for the year 2030 varying by a factor of 7 – despite comparable assumptions of economic and population growth! Energy consumption is evidently not a matter of fate, but political resolution. In other words, energy policy suffers not so much from a technology deficit, but rather and above all from a policy and implementation deficit. Energy policy is failing to exploit the existing technically realizable and macroeconomically cost-effective options and leeway for decision-making. Grave risks could be avoided, but too little is being done.

Even, for instance, in the IIASA "Low" Scenario, which was considered moderate at the time, the primary energy requirement rises to 22.4 TW by the year 2030; nuclear energy rises to 5.17 TW (=23%), and renewables to 2.28 TW (=10%). 67% of the primary energy requirement would then still have to be satisfied from fossil sources – with the consequence that CO_2 emissions would grow to almost twice the level of 1985 (to 9,400 million t C). This clearly shows that the earlier IIASA scenarios of the 1980s (in particular the "High" Scenario with 35.6 TW in 2030, of which 8.1 TW nuclear) have a *risk-cumulating effect.* Despite the exorbitant growth of nuclear energy, CO_2 emissions rise dramatically in both scenarios (in the "High" scenario three-fold between 1985 and 2030 to 15,800 million t C).

We can generalize this statement. All supply-oriented global energy scenarios – including those of the world energy conferences in Montreal (1989), Madrid (1992) and Tokyo (1995; for the exception see further below) have risk-cumulating effects: more CO_2 emissions (global warming), more nuclear energy (accident hazards) and more oil consumption (military conflicts)! Scenarios that attempt to resolve global energy problems only by means of a growing fossil or nuclear energy supply and an ever more complex diversification of the supply mix are funda-mentally incompatible with the goals of climate stabilization and risk minimization policy.

The basic fallacy of supply-oriented global energy scenarios is that they revolve around an extrapolation of substantial growth in primary energy consumption derived from the linkage of two key parameters – growing global population and excessive *per capita* energy consumption – instead of examining in detail the possibilities of enhanced energy efficiency in energy generation and energy conversion. In view, however, of the inexorable population growth in developing countries and their desire to raise their *per capita* living standard to that of the industrialized countries, *the global energy problem is first and foremost an efficiency problem, and at a second level a distribution problem.* The prime strategy must therefore be to cut drastically (by at least half) *per capita* consumption in the industrialized countries (through more efficient use), and to keep from the very start the unavoidable development-related growth in *per capita* consumption in the developing countries with a rising standard of living as small as possible by means of employing state-of-the-art energy conversion technologies. There is much evidence for our main thesis: without a substantial improvement in the efficiency with which every 'consumed' kilowatt-hour of energy is used[1] it will neither be possible to contain the global risks of climate change and of nuclear energy, nor to resolve peacefully the distributional problems of scarce oil and gas resources.

This pivotal realization has now also been brought home to the world's major conference of energy suppliers, the World Energy Council (WEC). At the last WEC conference in Tokyo (October 1995), a long-term scenario (to 2050 and 2100) was presented that – for the first time in the WEC context – intensively pursues the question of whether a risk-minimizing 'sustainable' energy strategy is globally feasible (WEC and IIASA, 1995). By WEC standards, the findings are sensational: in the 21st century, the challenging targets of a global climate protection policy can largely be achieved (CO_2 concentration rise less than 450 ppm, global temperature rise less than 2 °C above preindustrial levels) while simultaneously phasing out nuclear energy worldwide. An adequate CO_2 reduction (by about 2/3 compared to 1990) is, however, only achieved well after the year 2050. The German Climate Study Commission (Klima-Enquete) and the IPCC, however, demand a global reduction of CO_2 emission levels by 50% by the year 2050 (Enquete-Kommission, 1995; IPCC, 1995). The slower pace of the change in WEC Scenario C1 is of course primarily due to the assumed moderate average annual rise in energy productivity of 1.4%, although in 1992 the WEC still considered up to 2.4% to be feasible.

The risk-minimizing WEC Scenario C1 'coexists' in the new WEC/IIASA analyses with five further variants (A, B, C2) that display more or less risk-cumulating effects. Thus in Scenario A2 (A1), CO_2 emissions rise by a factor of 2.5 (2.0) and nuclear capacity is more than doubled (quadrupled) by the year 2050. The most important new findings of WEC are, however, to be seen in the appraisal of the investment costs and realization probabilities of the scenarios:

- For the period from 1990 to 2050, the cumulative investment costs of Scenario C1 are 33% and 43% lower than the corresponding investment costs of Scenarios B and A1! Customer-side investments have admittedly not been considered in any of the scenarios, and will presumably be higher in C1 than in the others due to the higher efficiency growth. On the other hand, these efficiency improvements also avoid more running energy costs for the customer. It thus appears justified to expect that the risk-minimizing Scenario C1 will also be the more beneficial from an economic perspective, too.

- The authors stress that despite the long time horizon, the basic decisions as to which path is to be taken must be made today, as the strategies modeled in the scenarios will become mutually exclusive within a few decades, and the capital tie-up periods in the energy system generally span several decades. Concerning the closeness to reality of the six scenarios, the authors

---

[1] Ernst Ulrich von Weizsäcker speaks of a feasible enhancement of energy productivity by at least a factor of 4 (Weizsäcker et al., 1995).

explicitly stress that "All are held to be realizable. But none assumes that the developments will occur of their own accord." (WEC/IIASA 1995, p. 2). There is therefore a need at all events for energy policy action. Risk minimization is possible and financable, but we – energy managers, politicians and the public at large – must all make up our minds soon.

1.2 Climate protection policies avoid grave future risks – and are cost-effective

In this context it is not possible to analyze the problems of global cost–benefit analyses in climate protection. In order, however, to round off the concept of the 'prevention economy' put forward here, the conclusion of the previous section needs to be extended by the damage dimension of climatic changes. The example of the WEC Scenario C1 has shown that a risk-minimizing strategy (climate protection plus nuclear phase-out) is already more cost-effective than conventional supply-oriented strategies, without consideration of the damages avoided by the former – while the latter are further associated with more risks and larger damage potentials (Grübler et al., 1995). This statement flies in the face of a school of global benefit-cost analyses whose best known and most influential representative is William Nordhaus (1993). This is not the place for a fundamental critique of the neoclassical general equilibrium models used by Nordhaus and others (see on this e.g. Hennicke and Becker, 1995 and the literature cited there). The dubiousness of Nordhaus' policy recommendations, however, shall be briefly summarized, based on sensitivity analyses using the DICE model. Nordhaus uses this model to argue that an "optimum climate protection policy" is one that applies at most a moderate tax level (7 $/t C) and is geared towards a low long-term CO_2 reduction rate (approx. 15%). This ultimately boils down to the recommendation that, by and large, 'adjustment' to climatic changes is the macroeconomically most cost-effective long-term strategy. As shown above, this is economically in direct contradiction to the WEC Scenario C1. But Nordhaus' own (most dubious) highly aggregated neoclassical model, too, can – with more plausible assumptions than Nordhaus uses – be applied to 'prove' the opposite result. If we base our calculation on an initial growth rate of energy productivity of 3% per year (reduced per decade by 11%; Nordhaus' initial rate being 1.25%) and a halving of the CO_2 intensity of the energy system by the year 2100 – which is realizable through more use of renewables – then the DICE model provides the following result: far from the long-term temperature rise of almost 6 °C held by Nordhaus to be "optimum" and acceptable, the recommended threshold value of climate protection policy (max. 2 °C) is not exceeded. Other reasonable parameter variations (e.g. low or no long-term discounting of climate damages, and a utility function that also takes immaterial damages into account) can also be used to make the modified DICE model justify – in contradiction to Nordhaus – a forceful ("optimum") CO_2 reduction rate corresponding to IPCC recommendations. Cost-benefit calculations of Nordhaus' kind and the consideration of climate damages thus do not call our recommendation for a forceful climate protection policy into question, but rather confirm it.

2. European Union: Risk minimization is financable

For the five largest European countries, the IPSEP study (see Krause et al., 1995) presents one of the most comprehensive scenario analyses, and an appraisal of the relative investment costs of a climate protection strategy. As the scenario results largely tally with the WEC Scenario C1 and studies undertaken for Germany, they need only be briefly reported here. Figure 2 gives an overview of the scenario results and the framework data assumed. In the two target scenarios, CO_2 reductions of almost 40% ("minimum costs") and almost 60% ("minimum risk") are achieved by the year 2020. The "minimum risk" scenario assumes that the available CO_2 reduction potentials provided by efficient energy use and generation are largely exploited. Both climate protection strategies generally give rise – this having been tested by extensive sensitivity analyses – to lower investment costs than the reference scenario, the "Conventional Wisdom Scenario" of the EU Commission.

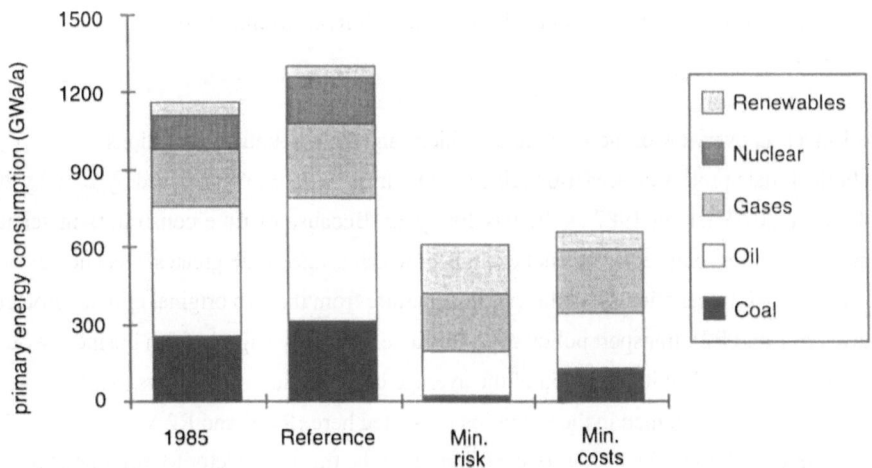

Figure 2. Primary energy use.

3. Germany: Room for manoeuvre and deficits in implementation

3.1 Business-as-usual versus climate protection and risk minimization: The scenarios of the German Bundestag's Climate Study Commission

The German Bundestag's Climate Study Commission has used the technique of scenario analysis to examine various future development pathways (Enquête-Kommission, 1995).[2] The findings of these scenario analyses could not be discussed in depth and agreed upon within the Study Commission process and are therefore not conclusive (see also the 1994 dissenting opinion ("Sondervotum") of the Social Democratic Party (SPD) in the Study Commission). Despite a multitude of unresolved questions and methodological problems, the scenarios nonetheless offer a more comprehensive basis for energy economy studies than has previously been available, and thus improved orientation for medium-term climate protection policies (Nitsch et al., 1990; Mashur and Brake, 1991; Traube, 1992; Hennicke, 1995).

The Study Commission scenarios first illustrate the development that is to be expected if current trends continue ("Reference Scenario"). The Reference Scenario illustrates that business-as-usual policies will fall far short of realizing environment and climate policy targets. Accordingly, two further scenarios were developed that lead to a CO_2 emission reduction of 45% by the year 2020 (with a 1987 baseline). The controversies surrounding the further use of nuclear energy were reflected by modeling two pathways. The first assumes nuclear capacity to remain constant over the medium-term (i.e. only replacement investments). The second assumes a phase-out of nuclear energy by the year 2005. In the following, these two scenario variants are referred to as "Climate Protection" and "Climate Protection with Risk Minimization".

Scenario definition

Table 1 gives an overview of the scenario definitions and main boundary conditions.

In both climate protection scenario variants, CO_2 emissions are to be reduced by 27% by 2005 and 45% by 2020, taking 1987 as the baseline year. Because of time constraints in scenario preparation, the transport sector was not examined, which is one of the greatest weaknesses of the Study Commission scenarios. Nonetheless, in departure from the two original climate protection scenarios (R1 and R2), transport policy measures diverging only slightly from business-as-usual (changes in the modal split, reduction of the average consumption of new cars to 5 1/100 km by the year 2020) were assumed in the scenarios presented here (R1 V and R2 V). However, these measures fall far short of the necessary contribution of the transport sector to formulated national

---

[2] The following section is based upon the article of Fischedick and Hennicke (1995).

Table 1. Scenario definitions and determining boundary conditions of the Study Commission scenarios

Scenario	Reference	Climate protection	Climate protection + risk minimization
Population development (millions) 1990: 79.8 2005: 81.1 (+ 0.1%/a) 2020: 79.1 (± 0.0%/a)			
Gross domestic product (1.000 million 1991 DM) 1990: 2 788.5 2005: 4 012.0 (+ 2.5%/a) 2020: 5 480.0 (+ 2.3%/a)			
Min. use of German hard coal (mill. tce)	50 (2005) 30 (2020)	45 (2005) 25 (2020)	45 (2005) 25 (2020)
Min. use of Eastern German brown coal (mill. t)	80 (2005) 80 (2020)	72 (2005) 40 (2020)	72 (2005) 40 (2020)
Min. use of Western German brown coal (mill. t)	106 (2005) 106 (2020)	95 (2005) 53 (2020)	95 (2005) 53 (2020)
Nuclear energy	constant capacity	constant capacity	phase-out by 2005
Transport sector	business-as-usual	measure mix	measure mix
CO_2 reduction target (baseline 1987)	no target	27% (2005) 45% (2020)	27% (2005) 45% (2020)

In the notation of the Study Commission, the Climate Protection Scenario is Scenario R1 V and the Climate Protection with Risk Minimization Scenario is Scenario R2V.

CO_2 reduction targets, so that all Study Commission scenarios assign an overproportional and thus also significantly more costly CO_2 reduction obligation to the energy sector. Moreover, the scenarios stipulate a minimum usage of German hard coal and eastern and western German brown coal.

Analysis of scenario findings

Figure 3 shows the resultant end-use energy consumption for the three scenarios.

If the developments assumed in the Reference Scenario come to be, then this will lead to a slight growth in end-use energy demand by the year 2020. Solid fuels increasingly lose importance. They are compensated for by gas and oil, whose aggregate consumption rises. Similarly, the demand for electric energy grows by an annual average of 0.6%. By contrast, the climate protection scenarios show a substantial reduction of end-use energy demand. Although in contradiction to theoretical and empirical arguments, it was assumed in the latter two scenarios that the

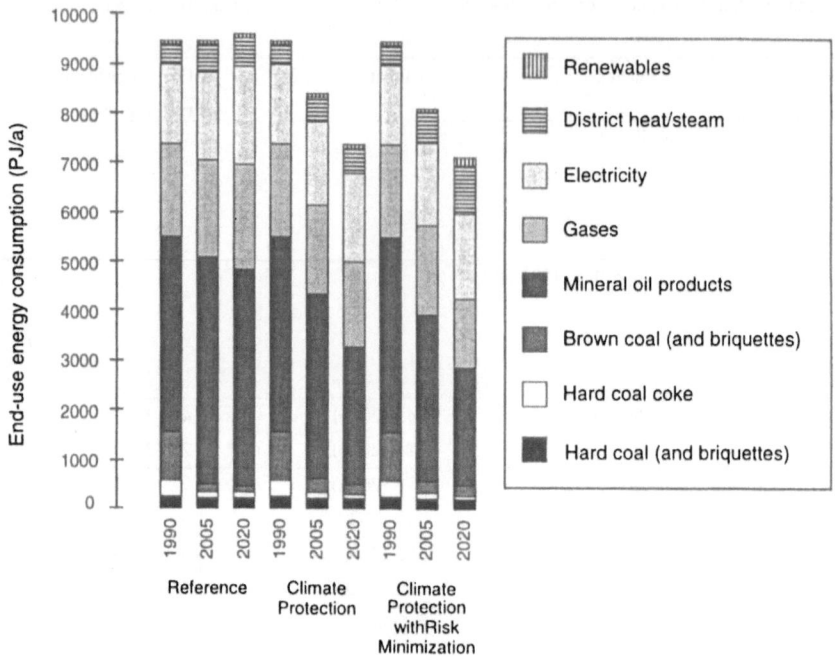

Figure 3. End-use energy consumption in the three Study Commission scenarios.

supply orientation immanent in a vertically integrated energy system relying on large-scale and nuclear power plants does not present any barriers to the exploitation of existing savings potentials. On this basis, and with a maintenance of present nuclear capacity, end-use energy demand drops by the year 2020 by some 22% from the 1990 baseline, and by about 24% compared to the reference development. In the third scenario, with a phase-out of nuclear energy, the end-use energy consumption reductions are not significantly larger. The selected model structure (linear programming model) and the assumed excessive costs of electricity-saving technologies nullify the phase-out-induced innovation and investment thrust indicated, for instance, in the dynamic input-output analysis of the ISI/DIW study conducted on behalf of the Study Commission.

The main part of the savings is due to a reduction in gas and mineral oil consumption, while in the climate protection scenarios, too, electricity demand rises slightly from the 1990 baseline. Renewable sources of energy find greater application in the satisfaction of end-use energy demand, particularly through the use of alternative fuels in the transport sector. Proceeding from the development of end-use energy demand outlined above, the associated primary energy consumption and resulting CO_2 emissions follow as shown in Figures 4 and 5.

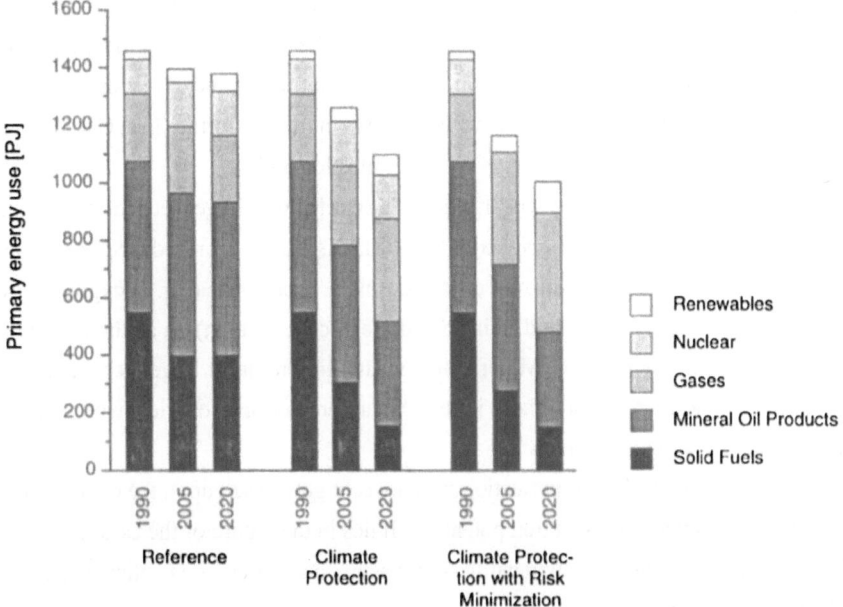

Figure 4. Primary energy consumption in the three Study Commission scenarios.

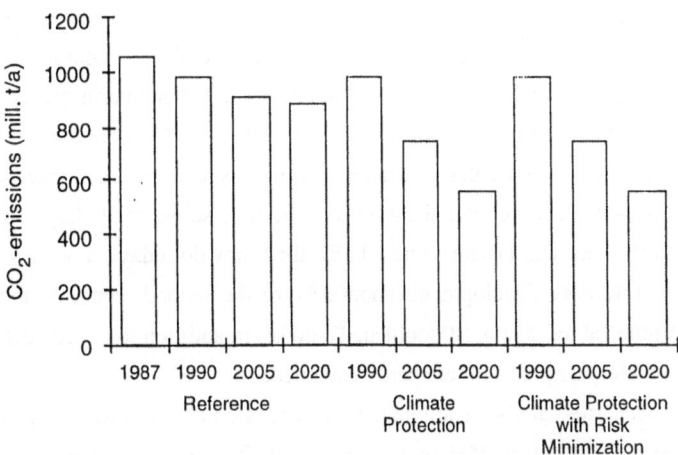

Figure 5. CO$_2$ emissions in the three Study Commission scenarios.

Primary energy consumption drops in all three scenarios, but to substantially varying extents. With the reference development, consumption drops from the 1990 baseline by 8.7% by the year

2020, leading to a reduction in CO_2 emissions of 14% by 2005, and 16% by 2020. This means that today's politically stipulated reduction targets would be far from being fulfilled with a business-as-usual development – the consumption reduction is mainly due to the restructuring of the (brown coal) energy system and general industrial restructuring in former East Germany, and can only marginally be viewed as a success of climate protection policy.

By contrast, the climate protection scenarios show a significantly larger reduction of primary energy consumption, namely by 24.3% and 30.6%. This is based upon reduced energy consumption on the demand side, and improved efficiency on the supply side (e.g. through improved average conversion efficiency ratios, and an increased use of cogeneration). In all three scenarios, the consumption of solid fuels drops. Mineral oil consumption remains largely constant in the Reference Scenario but drops substantially in the climate protection scenarios, while gas consumption rises considerably in the latter.

The analyses show that a climate protection strategy is largely based upon the exploitation of the existing conservation and substitution potentials. It lies in the nature of the assumptions and the model used that this applies largely regardless of whether nuclear energy continues to be used or not. The two climate protection scenarios are characterized by an average annual efficiency enhancement on the end-use side of 3.1% throughout the period from 1990 to 2020. By contrast, the reference development only entails an annual end-use efficiency enhancement of some 2.2%. When compared to the actual figures of the last 20 years, in which period the average annual efficiency improvement was 1.75% in former West Germany, these scenario figures imply substantially intensified conservation efforts. In the third, risk-minimized scenario, the compensation for the nuclear share of power supply is provided by an increased use of gas for an interim period in the conversion sector, an exploitation of primary savings potentials through the increased use of cogeneration, and an increased use of renewables. The latter, however, is relatively moderate. Renewable sources of energy do however account in the third scenario for a substantially larger proportion of energy supply in the year 2020 (with some 10.5%) than they do today (about 2% in 1990), but compared to the reference development (about 5%) or the second scenario (about 6.3%), the contribution of renewables is at most doubled. To do so, in addition to hydroelectric power the available wind and biomass potentials are partly exploited.

Both climate protection scenarios satisfy the stipulated CO_2 emission targets: CO_2 emissions are reduced by 27% by the year 2005, and 45% by the year 2020. On the basis of the selected assumptions (low CO_2 reduction contribution of the transport sector; excessive costs of energy savings; relatively favorable cost estimates for nuclear energy), the second scenario with a maintained nuclear capacity indicates lower macroeconomic expenditures than the reference development. By contrast, the third scenario with a simultaneous phase-out of nuclear energy indicates – again on the basis of these assumptions – cumulated, discounted extra costs over the period from

1990 to 2020 of some 150,000 million DM (or 5,000 million DM/a, i.e. about 60 DM *per capita* and year) compared to a development path following current trends. Compared to the second scenario with further use of nuclear energy, the assumptions favoring nuclear energy lead to extra costs of some 180,000 million DM, or about 75 DM *per capita* and year. We may assume for one thing that such a *per capita* burden would appear acceptable to most of the public as an 'insurance premium' for the avoidance of the risks of nuclear energy. For another, a more detailed analysis of the findings, taking into account the inconsistencies in the methodology and assumptions, the insufficient modeling of CO_2 reduction options in the transport sector and the uncertainties in the price development of energy carriers, would come to the conclusion that a phase-out of nuclear energy in conjunction with a fulfillment of CO_2 reduction targets can be achieved entirely without additional macroeconomic burdens. Indeed, similarly detailed studies conducted elsewhere, in which comprehensive comparisons of supply- and demand-side resources were calculated within an integrated examination of energy supply and demand, come to the conclusion for Germany that a nuclear phase-out undertaken within the framework of effective climate protection measures need not lead to extra costs, but can even lead to reduced costs and thus positive macroeconomic effects (see Enquete-Kommission, 1995:441; Krause et al., 1995b).

3.2 Instruments of climate protection and risk minimization policy

The scenarios presented above have shown – despite all their methodological deficiencies – that an adequate pool of technological options is at hand to provide an energy supply that is climatically acceptable and entails a minimum of risks. The crucial question, however, concerns the issue of which instruments can be applied to break down the barriers impeding climate protection and risk minimization policies, and thus remove the innovation and investment blockages currently standing in the way of a soft energy path. What is needed is a sector and target group-specific policy mix that does justice to the exigencies of climate protection.

Allocative processes in the market economy that are directed by competition and the private exploitation of capital have until now forced an astounding economic dynamism and – as long as economies and markets have *grown* – a certain degree of efficiency. But these 'acquisition techniques of the market' can neither set global quality and reduction targets, nor do they have self-regulating mechanisms that may solve distributional issues between rich and poor or between countries or generations. The German Climate Study Commission has shown beyond doubt that the challenging CO_2 reduction targets set in Germany (30% by 2005, 80% by 2050) can only be achieved by a *comprehensive and differentiated policy mix*. Strong political guidance must be given and decisions must be taken that put developments on the right track if climatically acceptable future markets are to be tapped in time and to a sufficient extent. The *concurrent shrinking of risk-burdened markets* (for fossil or politically unacceptable nuclear energy carriers) *and the politically induced emergence of 'soft' markets* (e.g. for energy-efficient universal technologies on the user side – in electric drives, ventilation, pressurized air, lighting; for renewable sources of energy; for low-energy houses; or for combined heat and power generation) demand long-term planning security for investors – something that can only be created through framework conditions set by the state. The guiding ideas of economic policy for a new *combined regime of self-control and state regulation* in the energy system are the 'economy of prevention', 'efficiency revolution' and 'new models of wealth'. The fundamental idea of a benefit-driven energy-efficient economy is that the prevention of unnecessary consumption of energy and resources must not only provide benefit to the customers, but also to the suppliers. Failure to give the profit-orientation of individual economic actors a new, ecological market and development perspective must leave ecology and economy incompatible in a profit-driven market economy, and objectives such as climate protection and sustainability would remain unachievable in market economy systems.

Take the example of energy taxation (as a first module of ecological tax reform). If well designed, this market economy demand management instrument acting via relative prices ('bads' becoming more expensive while 'goods' become cheaper) can accelerate the ecologically necessary structural change in a manner that is more conducive to a thriving economy and social equity than any laissez-faire policies. *Macroeconomic* studies provide further evidence that a very gradually rising and revenue-neutral energy tax will tend to trigger positive effects upon the overall economy (more employment, more qualitative growth). Not a single macroeconomic study to date (11/95) has come to a contrary conclusion – what we do find, however, is vociferous lobbying from the small camp of 'losers', and too little political support from the majority camp of 'winners'. To be moved by such lobbying to abstain from the introduction of an energy tax would be macroeconomically irrational. As long as no OECD-wide introduction of such a tax is on the horizon, it does remain purposeful, though, to use limited exemptions or transitional provisions to cushion the adjustment losses inevitably experienced by individual industries in any structural change (see Görres et al., 1994).

That these positive macroeconomic model results point in the right direction is verified by empirical evidence, and also by an overwhelming body of potential studies. The empirical evidence shows that the growth rates of demand-side and supply-side energy efficiency products have consistently been about twice as high as those of manufacturing over the past 10 years (see Schmidt, 1995). In the 1990s, this lead has lengthened. The German Federal Environmental Agency (UBA) estimates the total general expenditures for environmental protection in the period from 1975 to 1991 to have been 420,000 million DM. The jobs provided by environmental protection are estimated at 680,000 (1990), and the UBA estimates that this figure will increase to 1.1

million by the year 2000. As the UBA quite rightly concludes, "high environmental standards are the locational advantages of tomorrow". Without proactive environmental policy interventions in the markets, these growth markets would not have emerged, nor would Germany be by far the largest exporting country for environmentally-relevant goods, with a world market share of 21% (1990).

Present environmental protection expenditure (44,000 million DM in 1993) is, however, still largely made up of end-of-pipe technologies, which lead to new markets and returns for the manufacturers, but generally extra costs for the users (e.g. flue gas purification equipment for power plants). An 'economy of prevention', by contrast, relies primarily upon 'integrated pollution control', i.e. upon technical innovations that size and control processes and products in such a way that the user, too, receives substance, material and energy savings and cost savings in general, and thus the basis for an improved competitive position.

This explains the great *economic policy* relevance of improved energy efficiency in conversion and use. It is certain that almost 50% of Germany's energy consumption could be saved with the technologies that are already known today. With today's energy price level, this could cut the macroeconomic energy bill by some 100,000 million DM per year. The tapping of this savings potential could create half a million secure jobs (net after deduction of losses suffered in the energy supply sector). There is no doubt among energy experts as to the technical feasibility of this veritable 'efficiency revolution'.

But only a proactive policy – the rediscovery and active exercise of the primacy of energy policy – can bridge the widening gap between knowledge and action. Unlike on the energy supply side, frameworks must be judiciously set out and vigorously supported by all agencies of the state, if the energy consumers are not to dictate a much too high future energy consumption through their daily undertaking of thousands of energy-inefficient investments (Hennicke et al., 1994).

In contrast to the highly powered supply side, energy conservation has no lobby. Efficiency manufacturers, NGOs and the body politic as yet only have a marginal pull on the market for energy services. Investments in more energy supply are calculated with payback periods of 15 and more years, while energy-saving investments must pay back from the perspective of the users within 5 years at most. As a consequence, without state intervention there is a steady flow of too much macroeconomic capital into the expansion of supply. A full-scale study prepared for the Stadtwerke Hannover (1995) utility by the Öko-Institut and Wuppertal-Institut has shown for the exemplary case of this large municipally owned utility that in Germany, too, about 30% of electricity consumption could in principle be economically saved if the barriers impeding an operable 'competition of substitute services' between energy (electricity) and capital (end-use efficiency technologies) were removed. One means to this end is the consistent application of the concept of Least-Cost Planning, implying that the construction of new generating capacity is only authorized if no cheaper electricity savings potentials can be tapped. Extrapolated for the whole of Germany, over a period of 10 years LCP electricity conservation programs could save a good 20% (some 18,000 megawatts) of generating capacity, and be financed with a moderate rate increase of 1–2 Pf/kWh (including an additional profit as incentive for the utilities) – the annual macroeconomic electricity bill would still drop by 10,000 million DM (Hennicke and Seifried, 1994). Despite a slight rise in rates in order to finance the conservation programs, the aggregate customer bill for electricity and efficiency technologies would drop. And it is after all energy bills that ultimately count for the competitiveness of industry and the pocket of the individual – not the energy rates!

The 'building of conservation power plants' through LCP energy efficiency programs of utilities thus has high economic and ecological benefits. But without appropriate flanking regulatory policies, such 'conservation power plants' have little chance of succeeding in the struggle against the prevailing perverse incentive structure: At present, the more energy a utility sells and thus also damages the environment, the higher its profit is. The opposite could and should be the case: earning more with less energy. This is the applied 'economy of prevention'. A targeted removal of barriers in conjunction with a policy mix comprising energy taxation, the substitution of the anachronistic Energy Industry Act ("Energiewirtschaftsgesetz, EnWG of 1935") by a modern Energy Conservation Act, incentive regulation upon the basis of LCP/IRP (Integrated Resource Planning), contracting, standards (such as the German Heat Utilization Ordinance, "Wärmenutzungsverordnung") and grant programs can be applied to make energy conservation profitable – not only for efficiency manufacturers and customers, but also for the former 'energy supply companies', now the 'energy service companies of the future', and new players on the efficiency markets (e.g. energy agencies).

In principle, the 'economy of prevention' also functions on the heat and primary energy markets, and in support of structural change. Here there is a particular need, however, for control and flanking support by economic policy, in order to cushion the negative side-effects of vigorous ecological structural change. The market volume for the energetic upgrading of the building stock in Germany figures some 200,000 million DM alone, and to this come a further 400,000 million DM for 'routine' replacement investments. In cooperation with partners, a multi-market company such as the Ruhrkohle AG, which already has extensive experience in building management, could play a leading role on this market. Such intensified diversification activities of a formerly pure mining company could compensate for a part of the jobs no longer tenable in mining for reasons of climate protection.

The implementation of the 'economy of prevention' and a risk-minimizing energy policy must be linked with measures geared to removing barriers, and a sector and target-group specific mix of instruments. From the great array of available measures and instruments, we list some of the more important ones (see also Krause et al., 1995a):

- Accelerated transformation of utilities from energy supply companies to energy service companies
- Impulse program to create an electricity-saving infrastructure, following the Swiss model ("Rationelle und wirtschaftliche Nutzung von Elektrizität", RAWINE, the Swiss scheme for the efficient and economic utilization of electricity)
- Appliance/equipment standards and mandatory electricity consumption labeling
- Expansion of energy advice services, and promotion of the foundation of energy agencies
- Establishment of 'Round Tables' and 'Energy Councils'
- Coordination of work on efficiency programs
- Inclusion of decentralised CHP facilities in regulations governing feed-in to the electric grid
- Promotion and market introduction of renewable sources of energy (tapping of pilot markets) and intensified research and development efforts
- Substitution of the present national electricity tariff code ("Bundestarifordnung Elektrizität") by a general electricity price code
- Amendment of the national ordinance governing franchise fees ("Konzessionsabgabenverordnung"), and decoupling of energy revenues from local authority finances and the financing of local public transport
- Adoption and implementation of the German Heat Utilization Ordinance ("Wärmenutzungsverordnung")
- Formulation and implementation of energy conservation action plans in firms
- Promotion of performance contracting

4. Outlook

The above scenario analysis refers to the period to the year 2020. The selection of instruments targets an even shorter, medium-term time horizon. Accordingly, the measures for further reduction of energy consumption and associated environmental impacts assumed in the development pathways that appear necessary by the middle of the next century can only be viewed as first steps. If we are to take seriously all the risks perceptible today in the global energy system, then for the future we arrive not only at the necessity of more far-reaching reductions in CO_2 emissions, but generally of a transition to a low-risk and durable energy system. This means above all an intensified use of renewables to satisfy energy demand, in addition to the complete exploitation of available energy savings potentials and the quest for new, environmentally sounder forms of production and life styles.

It is at present unclear whether recourse to the potentials available in Germany will largely suffice, or whether an import of e.g. solar power from the southern countries of Europe (where solar irradiation is much higher and the use of solar thermal power plants thus possible) is the macroeconomically more efficient alternative – or indeed the inevitable alternative for the energy system of the future. The vision of a 'hydrogen economy' – which is often given far too much credence – becomes a very distant prospect against this background. Hydrogen will in future above all play a role as a storage medium for the balancing of energy supply and demand at the regional level, but a transport of hydrogen over great distances (e.g. from the North African countries) is largely out of the question (see Hennicke, 1995). A number of long-term studies conducted for the former West Germany have shown that a 70–80% reduction in CO_2 emissions is possible by the middle of the next century (Nitsch et al., 1990; Mashur and Brake, 1991; Traube, 1992). In view of the scarcity of capital and resources, this requires a careful selection process between the various technical options. No decision should be taken today that impede or even render impossible the achievement of climate protection targets in the future. In particular, investments in structures fundamentally at odds with climate protection objectives (e.g. traffic-intensive settlement structures; excessive levels of coal-fired power generation) must be avoided.

In a long-term systems analysis, the energy system cannot be viewed in isolation from the other sectors of the economy. The analysis must rather examine the overall system, including all sectors of economy and consumption. This entails the necessity of identifying and evaluating the various positive and negative sectoral interactions – something that the long-term studies conducted to date have only attempted to a limited extent. The design of a sustainable economic system must proceed from this basis with the formulation of a policy mix that aims, in addition to the reduction demands upon the energy system that are required from the present-day perspective, at a reduction of the consumption of 'environment' in general (energy, land use, materials consumption and the derived environmental impacts), and of all risks associated with energy consumption (BUND and Misereor, 1994). The interactions arising between the individual measures instituted to achieve the various reduction targets or requirements are multitudinous. Thus a reduction of heat consumption leads on the other hand to an increase in material consumption. Similarly, the increased utilization of renewable sources of energy (e.g. photovoltaics for power) leads to increased consumption of materials and land. This is further compounded by a multitude of competing utilization options (e.g. crops for energetic utilization or for raw materials production).

This background in particular makes it clear that a development pathway that is to lead to far-reaching reduction in the environmental impacts associated with energy supply cannot, over the long-term and with growing wealth expectations of the world's population, be based exclusively upon purely technical measures, e.g. the exploitation of savings potentials and the intensified use of renewables. Such a pathway must rather involve social innovations and societal structural changes leading to reduced demand for energy services. Thus Norgard, for instance, comes to the conclusion in his study of 15 European countries (Norgard and Viegand, 1992) that the exploitation of all demand-side efficiency improvements that are cost-effective or only involve slight additional expenditure by the year 2010 can indeed lead to a halving of electricity demand compared to 1988 levels, but that the demand for electric energy will rise again after 2010 if the growth in *per capita* demand for electricity-specific energy services cannot be contained. If, however, such containment succeeds, then Norgard shows that the remaining electricity demand in Western Europe can be entirely covered from renewable sources.

Particularly in view of the sharpening distributional problems (in industrialized countries as well) it cannot be our concern to define certain 'sustainable' *per capita* consumption levels. Instead, a process of societal research and quest must be organized, seeking for ways in which dematerialized and de-energized services and new models of wealth can generally contribute to an attenuation of unnecessary and environmentally damaging throwaway consumption. Catchphrases for such approaches are, e.g., closed-loop economy and recycling, leasing ("use rather than own"), durability, multiple use and reparability of products, exchange communities, flat sharing, car sharing, neighborhood support structures, new forms of living together, communal day-nurseries and launderettes. Furthermore, a greater degree of municipalization and decentralization of economic structures and infrastructure and social life in general can provide substantial reductions in consumption (e.g. through avoided traffic).

The drafting of such total-system scenarios that include aspects of social change and take competing or interacting reduction requirements into consideration requires more detailed research, and should come to the forefront of energy policy discussion processes.

References

BUND and Misereor (eds) (1994) *Sustainable Germany*. Birkhäuser, Basel.
Enquete-Kommission Schutz der Erdatmosphäre des 12. Deutschen Bundestages (1995) *Mehr Zukunft für die Erde – Nachhaltige Energiepolitik für dauerhaften Klimaschutz*. Bundestagsdrucksache 12/8600, Bonn.
Fischedick, M and Hennicke, P (1995) Für eine Klimaverträgliche und risikominimierende Energieversorgung. *In:* Greenpeace (eds): *Der Preis der Energie*. C.H. Beck, München.
Goldemberg, J, Johanssons, TB, Reddy, AKN and Williams, RH (1988) *Energy for a Sustainable World*. TERI, New Delhi.
Görres, A, Ehringhaus, H, and von Weizsäcker, EU (1994) *Der Weg zur ökologischen Steuerreform. Weniger Umweltbelastung und mehr Beschäftigung*. Das Memorandum des Fördervereins ökologische Steuerreform, München.
Greenpeace (1993) *Towards a Fossil Free Energy Future. The Next Energy Transition*. FFES des Boston Center des Stockholm Environment Institute, Amsterdam.

Grübler, A, Jefferson, M and Nakičenovič, N (1995) *A summary of the joint IIASA and WEC study on long-term energy perspectives*. WP-95-102, IIASA, Laxenburg.

Häfele, W, Anderer, J, McDonald, A and Nakičenovič, N (1981) *Energy in a Finite World*. Ballinger, Cambridge.

Hennicke, P (ed) (1995) *Solarwasserstoff – Energieträger der Zukunft*. Birkhäuser, Basel.

Hennicke, P and Becker, R (1995) Ist Anpassen billiger als Vermeiden? *In:* P Hennicke (ed) *Klimaschutz: Die Bedeutung von Kosten-Nutzen-Analysen*. Birkhäuser, Basel.

Hennicke, P and Seifried, D (1994) *Endbericht "Least-Cost Planning" im Auftrag der "Gruppe Energie 2010"*. Wuppertal, Freiburg.

Hennicke, P, Richter, K and Schlegelmilch, K (1994) *Nutzen und Kosten von Energiesparmaßnahmen. Vorschläge für neue Förderinstrumente*. Studie im Auftrag der deutschen Ausgleichsbank, Wuppertal.

IPCC (Intergovernmental Panel on Climate Change) (1995) *Climate Change 1994: Radiative forcing of climate change and an evaluation of the IPCC IS92 emission scenarios*. Cambridge University Press, Cambridge.

Johannson, TB, Kelly, H, Reddy, AKN and Williams, RH (1993) *Renewable Energy: Sources for Fuels and Electricity*. Earthscan, London.

Krause, F, Koomey, J and Bach, W (1995a) Energy Policy in the Greenhouse. Vol. II. *Cutting Carbon Emissions: Burden or Benefit? The Economics of Energy Taxes and regulatory Reforms on Climate, Growth and Jobs. Executive Summary*. IPSEP, El Cerrito.

Krause, F , Koomey, J and Samstad, A (1995b) Energy policy in the Greenhouse. Vol. II. 3B. *Negawatt Power. The Cost and Potential of Electrical Efficiency Resources in Western Europe*. IPSEP, El Cerrito.

Lovins, AB, Lovins, LH, Krause, F and Bach, W (1983) *Wirtschaftlichster Energieeinsatz*. C.F. Müller, Karlsruhe.

Mashur, KP and Brake, H (1991) *Konsistenzprüfung einer denkbaren zukünftigen Wasserstoffwirtschaft*. Prognos AG, Basel.

Müller, M and Hennicke, P (1995) *Mehr Wohlstand mit weniger Energie*. Wissentschaftliche Buchgemeinschaft, Darmstadt.

Nitsch, J (lead author) et al. (1990) *Bedingungen und Folgen von Aufbaustrategien für eine solare Wasserstoffwirtschaft*. Bericht der Enquete-Kommission Technikfolgenabschätzung, Bonn.

Nordhaus, W (1993) Optimal greenhouse-gas reductions and tax policy in the "DICE"-model. *Am. Econ. Rev.* 83(2):313–317.

Norgard, J and Viegand, J (1992) *Low Electricity Europe – Sustainable Options*. European Environmental Bureau, Brussels.

Schüssler, M and Hennicke, P (1994) *Potentiale und Kosten für eine risikoarme Energieversorgung: Übersicht über die Ergebnisse internationaler Studien*. WP Nr.11. Wuppertal Institut, Wuppertal.

Schmidt, A (1995) *Ökonomische Auswirkungen rationeller Energieverwendung und erneuerbarer Energiequellen in den Bereichen Produktion und Außenhandel in Deutschland, 1976–1993*. (Diplom-Arbeit) Technische Hochschule Darmstadt, Darmstadt.

Stadtwerke Hannover AG (ed) (1995) *Integrierte Ressourcenplanung. Die LCP-Fallstudie der Stadtwerke Hannover AG (The Hanover Case Study)*. Wuppertal, Freiburg.

Traube, K (1992) *Perspektiven der Umstrukturierung des Westdeutschen Energiesystems angesichts des CO_2 Problems*. Bremer Energie-Institut, Bremen.

von Weizsäcker, EU, Lovins, AB and Lovins, HL (1995) *Faktor 4. Doppelter Wohlstand – halbierter Naturverbrauch*. Droemer-Knaur, München.

WEC and IIASA (World Energy Council and International Institute for Applied Systems Analysis) (1995) *Global Energy Perspectives to 2050 and Beyond*. WEC, London.

The value of advanced energy technologies in stabilizing atmospheric CO_2

Jae Edmonds and Marshall Wise

Batelle Pacific Northwest National Laboratories, 901 D Street, S.W., Suite 900, Washington DC 20024-2115, USA

1. Introduction

In the wake of the Berlin Mandate (COP, 1995), there has been considerable interest in emissions trajectories that would stabilize the atmospheric concentration of greenhouse gases, the objective of the Framework Convention on Climate Change (FCCC; United Nations, 1992). In light of its long lifetime in the atmosphere, and complex mechanisms which govern its concentration (IPCC, 1990, 1992, 1995), carbon dioxide (CO_2) is of particular interest.

The concentration of atmospheric CO_2 has risen from preindustrial levels of approximately 275 ppmv (year 1750) to 355 ppmv in 1990 (IPCC,1992). This rise has been closely paralleled by anthropogenic emissions, principally from land-use change and fossil fuel combustion. Strong circumstantial evidence exists to link the atmospheric build-up in concentrations with anthropogenic emissions (IPCC, 1992).

While the ultimate goal of the FCCC is the "stabilization of greenhouse gas concentrations in the atmosphere at a level that would prevent dangerous anthropogenic interference with the climate system," the FCCC is not specific about what constitutes "dangerous anthropogenic interference with the climate system." That issue also lies beyond the bounds of this paper. Rather, the purpose of this paper is to examine the implications of establishing various ceilings for atmospheric concentrations of CO_2.

The approach of this paper is to consider first some of the economic implications of stabilizing the concentration of atmospheric CO_2 and long-term emissions requirements, to maintain that concentration indefinitely. We then consider the economics of stabilizing the atmosphere at 550 ppmv and 450 ppmv. We next investigate the implications of accelerating the development and deployment of advanced energy technologies including their economic value. We have chosen to focus on concentrations of 550 ppmv and 450 ppmv not because there is any evidence that these ceilings are the maximum concentrations consistent with the FCCC, but rather because significantly higher concentration ceilings such as 650 ppmv and 750 ppmv, require fewer near-term actions to achieve and maintain. Alternatively, significantly lower ceilings such as 350 ppmv require immediate and dramatic actions to achieve. Concentration ceilings of 550 ppmv and 450 ppmv require immediate action to achieve and maintain, but their achievement can be either more or less expensive depending upon the emissions trajectory chosen.

2. The economics of CO_2 consideration ceilings

The identification of an emissions trajectory which simultaneously minimizes economic cost and satisfies an atmospheric concentration ceiling is a familiar problem to economists. Researchers in the area of climate change have examined variants of the problem for many years. Nordhaus (1979) was the first to examine the problem which has more recently been considered by Richels and Edmonds (1994), Kosobud et al. (1994), Manne and Richels (1995), and Wigley et al. (1996). Several reasons argue for a particular pattern of emissions if the economic burden to society is to be minimized. The generic path for an emissions trajectory which stabilizes emissions has three phases: 1. increasing emissions, 2. emissions peak, and 3. declining emissions. The extent of each of these phases depends on height of the ceiling, the cost of emissions reductions and the nature of the cost function, and on the carbon cycle.

The rationale for the three-phase optimal trajectory can be seen by considering a ceiling of 500 ppmv. Begin by noting that present fossil fuel emissions rates of 6 PgC/yr. imply a CO_2 concentration in year 2100 of 500 ppmv. Since constant emissions imply concentrations in the years beyond 2100 in excess of 500 ppmv, then observance of a ceiling implies that year 2100 fossil fuel emissions must fall below 6 PgC/yr. If this is the case, then emissions in some year prior to 2100 must be in excess of 6 PgC/yr., otherwise the cumulative emission condition would be violated. The economic and carbon cycle considerations listed below argue that any extra emissions be taken near the present:

1. *Carbon cycle*. Natural sinks have a longer period over which to remove emissions early in the period of analysis. This implies higher cumulative emissions for trajectories which stabilize at a similar concentration but with larger near-term loading. The carbon cycle dividend can be substantial. For the 550 ppmv computation the difference between cumulative emissions for the stable emissions and atmospheric stabilization cases is 60 PgC (10 years of fossil fuel carbon emissions at present rates).

2. *Discounting*. Any economic activity such as emissions reductions requires the allocation of scarce resources. A positive marginal product of capital means that resources set aside today can accommodate a larger burden, if they are not needed until some future date. Put another way, the further in the future a given economic burden lies, the smaller the resources that must be set aside today to facilitate the activity. There is great leverage to be gained by postponing

the burden of emissions reductions. This is not the same as procrastinating. There is still a present-day resource equivalent burden, but the overall resource requirements are lessened. Furthermore, it does not mean doing nothing. Emissions must eventually peak and decline and therefore the nature of new investments must change immediately, because each piece of new capital installation must spend an increasingly long period operating in a reduced emissions environment.

3. *Technology development.* Emissions reduction technologies are not static. As the IPCC Working Group II notes, multiple technologies exist in various stages of development and deployment which have the capacity to reduce emissions in the future at costs lower than those required today. Since cumulative emissions reductions are largely fixed (subject to the qualifications of the carbon cycle mentioned above) then costs are lowered by undertaking emissions reductions most strenuously when they are cheapest – in the future – since a ton of emissions reduction counts as much or more if it is undertaken in the future as it does if taken now.

4. *Capital utilization.* Both physical and human capital are created with a set of expectations about the environment in which they will be used. These expectations shape the character of these investments. Existing investments were made with no expectation of a carbon emission penalty. There is a cost to utilizing capital in an environment for which it was not configured. As a general rule, the greater the departure from the anticipated operating environment, the larger the penalty sustained. For atmospheric stabilization ceilings which do not require immediate emissions reductions, there is a benefit to announcing future emissions controls, but implementing them over time. See point 2 above for further discussion.

Richels and Edmonds (1994) have shown that the cost of efficiently stabilizing the atmosphere at 500 ppmv may be only half the cost of efficiently stabilizing emissions. But thinking about the problem of controlling the atmosphere has traditionally been focused on annual emissions rates. The Toronto (1988) goal of a 20% reduction in emissions[1] and the FCCC commitment of Annex I nations to return to 1990 emissions levels are examples.

But constant emissions rates eventually mean constantly rising concentrations. Framing the problem in terms of a static emissions rate is inconsistent with our present scientific understanding of the carbon cycle. If the CO$_2$ concentration ceiling is low, for example 400 ppmv or below, then emissions reductions must begin immediately and a 20% reduction in emissions is merely a point to be passed on a steadily declining emissions trajectory. There is no room for manoeuvre given an atmospheric concentration ceiling of 350 ppmv. For atmospheric CO$_2$ ceilings below

---

[1] The Toronto Climate Conference (1988) proposed a goal of reducing carbon dioxide (CO$_2$) emissions by 20% by the year 2005. This recommendation was established as a first step toward the eventual goal of reducing emissions to the point where atmospheric concentrations of greenhouse gases such as CO$_2$ would be stabilized.

350 ppmv, emissions must be negative in some years in order to stabilize concentrations within the next two centuries.

Hourcade and Chapuis (1994) have shown that the above arguments do not imply an indefinite period of postponement of emissions reductions. Waiting too long to initiate emissions reductions has an economic penalty in the same way as doing too much too soon.

While the generic character of the optimal emissions trajectory is well defined, the specific time path of emissions is at present uncertain. Grubb (1992,1995) and Hourcade and Chapuis (1994) have argued that if costs depend primarily on the rate of emissions reductions and not on the magnitude, then the shape of the curve should be flatter. They examine a case in which only transitions costs matter. By hypothesis, the cost of emissions reductions are zero in the long-term. Thus the problem becomes one of minimizing the rate of departure from the reference case. There is therefore an advantage to minimizing the maximum rate of departure. This in turn implies a relatively expeditious departure from the reference case and means that peak emissions are lower and near-term emissions reductions are greater. These considerations do not affect the fundamental conclusion of a three phase optimal emissions path.

3. Energy, agriculture, land-use, and economy

To begin to assess the potential implications for an agreement to stabilize CO_2 concentrations at 550 ppmv, we use the MiniCAM 2.0 model to construct a reference case which is similar in character to the scenario developed by the IPCC, IS92a (Leggett et al., 1992).

The MiniCAM 2.0 is an integrated assessment model with four major component parts: Human activities, Atmospheric Composition, Climate and Sea Level, and Ecological Systems. The model considers both energy and land-use change explicitly and interactively. The energy system model is the Edmonds-Reilly-Barns (ERB) model (Edmonds and Reilly, 1985), while the agriculture land-use model (ALM) was developed explicitly for MiniCAM 2.0. The approach to modeling atmospheric composition, climate change, sea level, and economic damages is the same as in MiniCAM 1.0 (Edmonds et al.,1994).

The ALM is described in Edmonds et al. (1995a). The ALM partitions land into managed and less managed systems. The managed lands are used intensively for human settlement and infrastructure and extensively for growing crops, raising livestock, managed forests, or biomass cultivation for energy use. Less managed lands are partitioned into ecological categories. The allocation of lands to human settlement and infrastructure is determined by population and income and takes precedence in the model over all other managed land-use. Extensive land uses are determined by expected profitability, which in turn depends on plant productivity, product prices, technology, fertilization, atmospheric CO_2 concentrations, climate, population, income and taxes, tariffs, and subsidies. Less managed lands include those which are "parked", that is excluded from use for managed activities, and those which are potentially available for managed uses.

The boundary between the managed and less managed systems is determined by the expected profitability of managed lands in general. Global markets are established for each of the major traded ALM commodities (crops, livestock, and forest products) and a world price is established which clears international markets. Biomass for energy use is determined interactively with the ERB. Since biomass is used as an energy resource, its demand and price is determined in the ERB while its supply is determined in the ALM. Changes in land allocations determine net biomass fluxes from the terrestrial biosphere, while the ERB determines energy related emissions. Other emissions, such as from cement manufacture and from chlorofluorocarbons (CFCs) and CFC substitutes are handled exogenously. Key assumptions used to reproduce the IPCC IS92a trajectory are reported in Table 1. Key reference case agriculture and land-use assumptions are given in Table 2. Note that reference case biomass assumptions imply a productivity of 17 Mg/ha/yr. by the year 2100.

While an enormous supply of energy is available in the form of shale oil, this resource is generally quite expensive to produce with current technologies and therefore is not a major factor

Table 1. Key Case 1 assumptions

PARAMETER	VALUE
Exogenous end-use energy intensity improvement rate	OECD: 0.5%/yr
	EEFSU [a]: 2.5%/yr
	CHINA: 2.2%/yr
	ROW [b]: 0.3%/yr
Fossil fuel resources	
Oil	18,011 EJ
Gas	17,451 EJ
Coal	271,000 EJ
Uranium	14,423 EJ (extend w. breeder option)
Conventional fossil fuel power plant efficiency	33% in 1990
	66% by 2050
Global population (year 2100)	$11,312 \times 10^6$
Global GNP growth	2.3%/year

[a] Eastern Europe and the Former Soviet Union.
[b] Rest of the World.

Table 2. Key Case 1 assumptions: Agriculture and land-use

PARAMETER	VALUE
Rate of exogenous productivity improvement	
Crops	1.0%/yr
Livestock	1.0%/yr
Managed forests	1.0%/yr
Fuelwood	1.0%/yr
Price elasticities	
Crops	−0.2
Livestock	−0.2
Managed forests	−0.4
Fuelwood	−0.4
Asymptotic demands for crops and livestock (Mg/cap/yr)	
Crops	0.95
Livestock	0.21
Income elasticities	
Managed forests	0.2
Fuelwood	−0.2
Biomass productivity (average potential in 1990)	6 Mg/ha/yr

in the reference scenario. The uranium resource base is assumed to be extended if a breeder reactor technology is adopted.

Energy and fossil fuel carbon emissions closely mirror those developed for IS92a. Global emissions are compared between IS92a and our reference case in Figure 1. Figure 2 compares regional energy consumption between IS92a and our reference case. Figure 3 shows the global energy consumption by fuel for these same cases. Both cases have been tuned to be generally consistent with the regional pattern of energy use and the modal pattern of energy use of IS92a.

In the reference case, energy system grows from its 1990 level of approximately 340 EJ/year to approximately 1440 EJ/year. Conventional oil and gas production peaks in the year 2020, and declines thereafter. Coal production grows steadily from approximately 95 EJ/year in 1990 to more than 720 EJ/year in 2095. Both biomass and solar electric[2] technologies show significant growth. By the year 2095, they provide 180 and 190 EJ/year, respectively.

---

[2] Solar electricity is a general category which includes all non-carbon emitting electricity technologies other than nuclear, hydro, and biomass. Thus fusion, wind, geothermal, and OTEC (ocean thermal energy conversion) are included in addition to photovoltaic and power tower technologies.

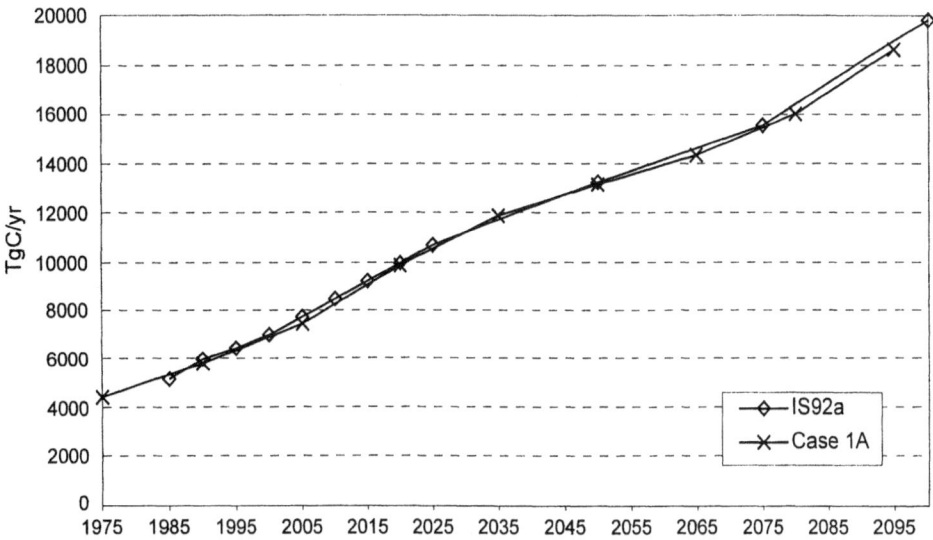

Figure 1. Global annual fossil fuel carbon emissions.

The regional distribution of fossil fuel CO_2 emissions changes greatly over the period of analysis. OECD regional emissions decline steadily as a share of global emissions. (They rise slightly in IS92a.) Emissions in the former Soviet Union and Eastern Europe grow, but only

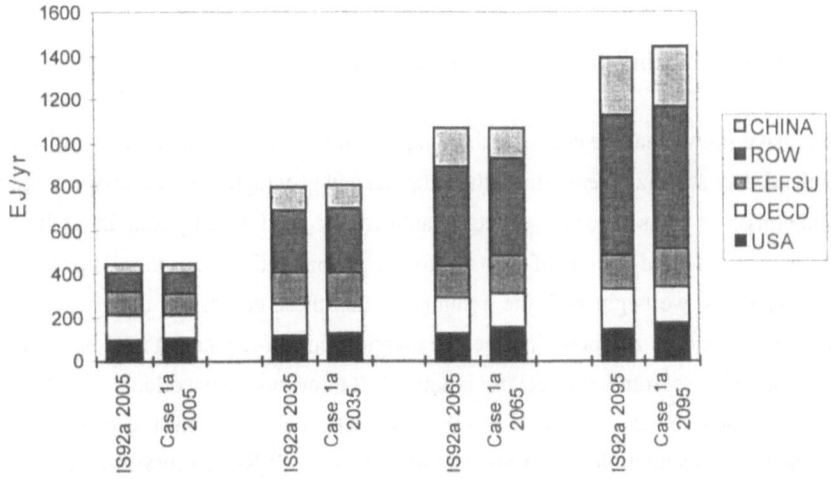

Figure 2. Primary energy consumption by region: IS92a vs. Case 1a.

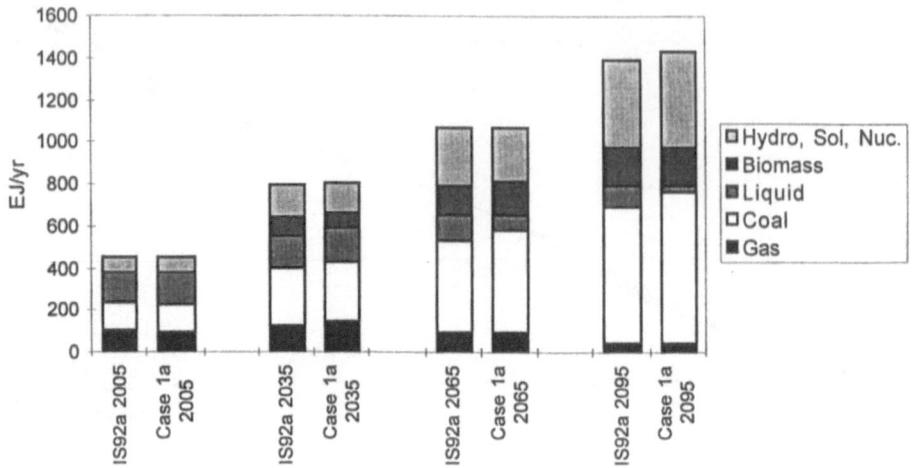

Figure 3. Global energy consumption, by fuel: IS92a vs. Case 1a.

slowly and this region's share of the global total erodes from a quarter to one fifth. Robust economic growth in the developing world fuels rapid emissions growth. China alone grows from 15 percent of global fossil fuel carbon emissions to 25 percent by the year 2095 in the reference case. The developing world as a whole increases its share from 30 percent in 1990 to 47 percent in 2095.

4. Implications of stabilizing the atmosphere at 550 ppmv

While others have shown that the cost of stabilizing the atmosphere is potentially significantly lower than the cost of stabilizing emissions, the temporal and geographic character of economically efficient emissions paths have not yet been examined. We begin here by examining the consequences for the world and nations of global efforts to stabilize the atmosphere. For the purposes of this analysis, we begin with the assumption that these efforts are undertaken in an efficient manner. That is, at any point in time, the marginal cost of reducing a net ton of carbon emitted to the atmosphere from fossil fuel use is equivalent throughout the world.

To accomplish the stabilization of the atmosphere, we deflect the emissions trajectory from the reference case by imposing a global tax so as to reproduce the WRE550 fossil fuel emissions trajectory given by Wigley et al. (1996). Tax rates for this case (WRE550) and other cases examined later are shown in Figure 4. While no taxes are required until the year 2020, tax rates

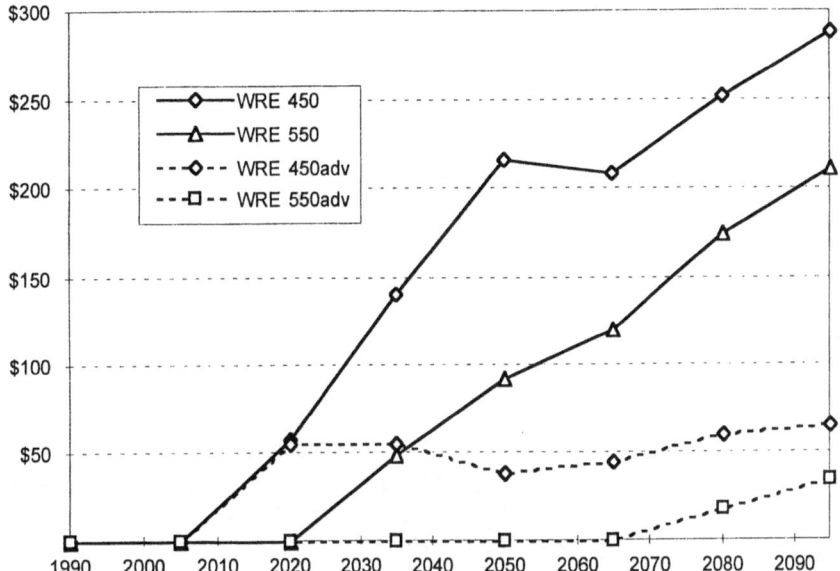

Figure 4. Carbon tax rate to accommodate an atmospheric CO_2 ceiling ($/MgC).

climb steadily thereafter until they reach a bit more than $200/TC in the year 2095. The economic burden, measured in terms of percentage loss in GDP, for four regions (OECD, China, Eastern Europe and the Former Soviet Union [EEFSU], and the Rest of the World) is shown in Figure 5. Burdens for the OECD and Rest of the World (ROW) regions rise slowly over time reaching about 0.5% of GDP by the end of the next century. Burdens for China and EEFSU are more substantial exceeding 1.0% before the end of the century.

To examine the question of Annex I leadership, we have computed the emissions reductions required for Annex I to control atmospheric CO_2 concentrations to the WRE550 trajectory assuming an efficient global system of joint implementation with appropriate side payments. We have plotted global emissions reduction obligations in terms of reductions in OECD emissions relative to 1990 levels in Figure 6 for a variety of emissions ceilings. Since global emissions reductions are trivial before the year 2020, Annex I emissions rise relative to 1990 levels until that date. By the year 2050, however, emissions reduction obligations begin to be substantial, 50% by that year. By the year 2065, emissions reduction obligations relative to 1990 emissions rates would have to exceed total emissions to track WRE550. This point is reached more than a decade earlier for the case WRE450.

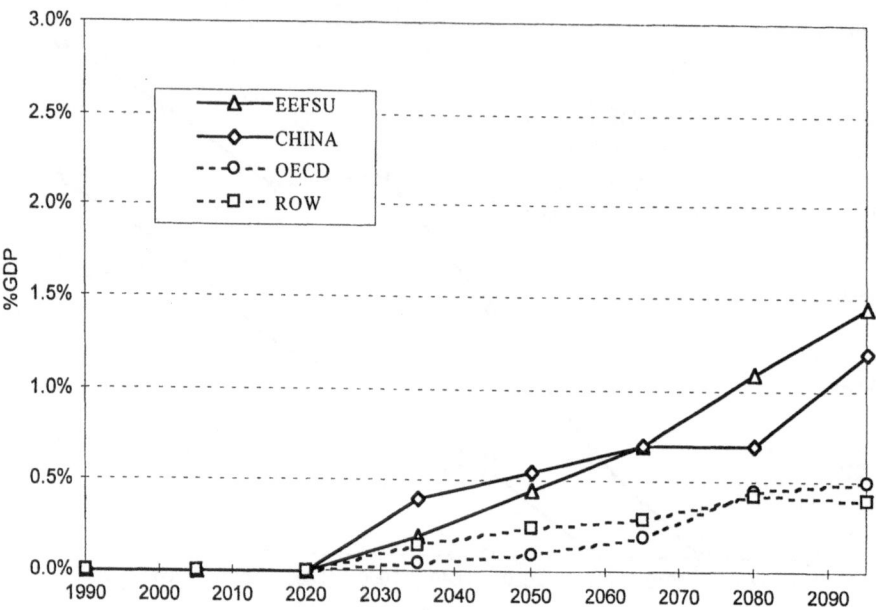

Figure 5a. Cost of stabilizing atmospheric CO_2 concentrations at 550 ppmv (%GDP).

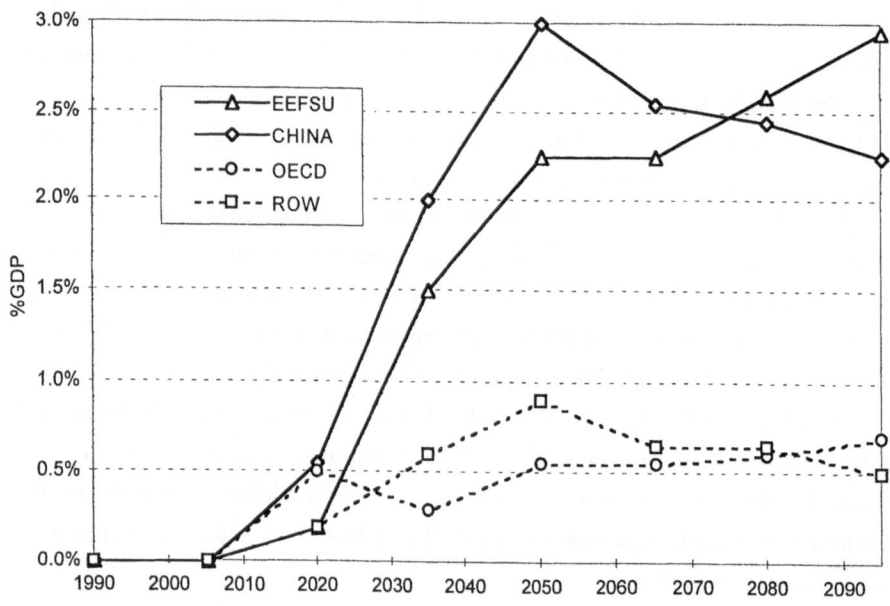

Figure 5b. Cost of stabilizing atmospheric CO_2 concentrations at 450 ppmv (%GDP).

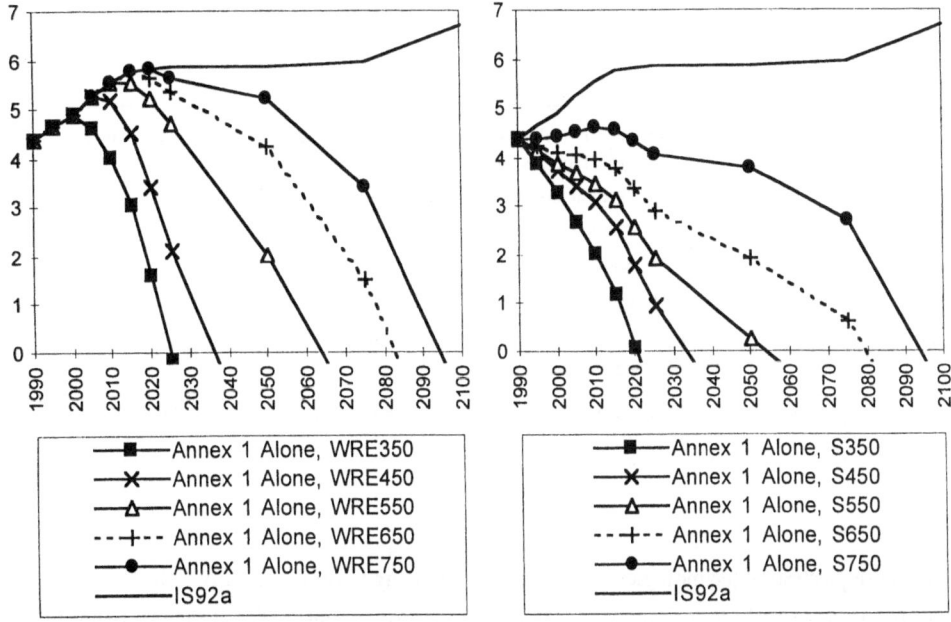

Figure 6. Annex I emissions trajectories necessary to affect WRE and S atmospheric stabilization scenarios on the assumption that non-Annex I emissions are unaffected. (PgC/yr).

5. The value of enhanced technology

There is ample reason to believe that the character, availability and usefulness of technologies for providing energy services can exert a profound effect on the cost of reducing emissions. Edmonds et al. (1994) explored the implications of the introduction of some advanced energy technologies and found that under some circumstances they would provide a means by which atmospheric concentrations of CO_2 could be stabilized. There are a variety of technologies which produce no direct emissions of carbon, including energy conservation, solar photovoltaics and other solar electric energy technologies, biomass energy[3] , nuclear fission, nuclear fusion, hydro-electric power, geothermal energy, wind energy, and tidal energy. The introduction of carbon cap-

---

[3] We note that while biomass energy releases carbon on combustion, it absorbs an equal amount in its growth phase. Thus over the course of its growth cycle biomass energy introduces no net emission of carbon. Several dynamic considerations enter into computing the overall effect. For example, if a biomass energy system is growing with planting in anticipation of energy use, there will be a growing stock of carbon in place and this stock will exceed annual emissions resulting in a net terrestrial sink. On the other hand, deforestation releases carbon which was removed from the atmosphere at a much earlier date. Harvest and oxidation of this stock results in net annual emissions. The overall dynamic impact of the introduction depends on the carbon intensity of the biomass energy system in comparison with the land-use which it displaced.

Table 3. Advanced energy technology assumptions

Technology	Description
Advanced liquefied hydrogen fuel cells	Hydrogen fuel cells are used to power transportation. Hydrogen is available from natural gas, biomass and electrolysis at the following costs:
	Natural gas steam reforming: $1.71 + P_{gas}/0.901$
	Biomass BCL gasifier: $4.83 + P_{biomass}/0.784$
	Electrolysis: $2.36 + P_{elec}/0.900$
	Hydrogen is assumed to be liquefied and used in fuel cells on board vehicles.
Solar, wind, and fusion power	The cost of solar, wind, and fusion power reaches a busbar cost of $0.04/kWh by 2020 and the cost decreases at 0.5 percent/year thereafter.
Biomass energy supply	By the year 2020, 20 percent of the biomass resource is available at $1.40/EJ, and 80 percent is available at $2.40/EJ.

ture and sequestration technologies holds potential for using fossil fuel energy technologies without atmospheric carbon emissions. Similarly, the improvement of fossil energy technology efficiencies reduces net carbon emission. IPCC (1996) Working Group II reports a variety of technologies that are either available or potentially available in the foreseeable future which could affect the cost of emissions reductions. We next consider the effect of the introduction of advanced energy technologies identified by the IPCC on the cost of stabilizing the atmospheric concentration of CO_2. These technologies are assumed to become available between 2005 and 2020 and improve at 0.5%/year. Technology assumptions are identified in Table 3.

We have conducted the same analyses as in the preceding section, but with the assumed availability of advanced energy technologies. This has a profound effect on the cost and burden of stabilizing the concentration of atmospheric CO_2. The marginal cost of carbon emissions reductions is zero until after the year 2050 when the atmosphere is stabilized at 550 ppmv (case WRE550, adv) and rises to only $40/TC by the end of the century. The costs of emissions reductions experienced by the world's regions are minor for the 550 ceiling (see Fig. 5a). Costs are higher with a 450 ppmv ceiling (see Fig. 5b). Emissions reductions relative to the reference case are required beginning in the year 2005, and advanced energy technologies are not assumed to be available until 2020. This produces near-term costs of emissions reductions, which for China exceed 0.5% per year between 2005 and 2050. The decline in emissions reductions requirements after the year 2035 leads to a decline in costs thereafter.

Emissions reduction requirements are far less strenuous when advanced energy technologies are available. This occurs because the reference case is dramatically altered. Reference emissions revised to include advanced energy technologies are shown in Figure 7. They peak in the year

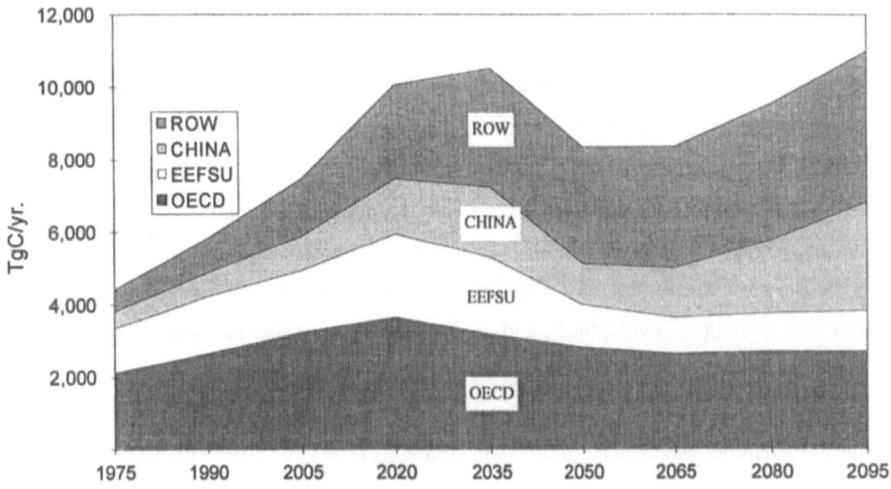

Figure 7. Fossil fuel carbon emissions with advanced energy technologies introduced in the year 2020.

2035, then decline, and then begin to rise again starting in 2065. The effect of advanced technologies is to change the non-tax trajectory into one which is similar, during much of the next century, to the WRE550 trajectory. Without intervention, the revised reference case IS92a, adv exhibits a concentration of 600 ppmv in the year 2100. Intervention to affect the emissions trajectory of WRE550 need not begin until after the year 2050. In this case, there is virtually no international transfer burden to be distributed. If technologies such as those described here can be developed and deployed by the year 2020, then the cost of achieving an atmospheric CO_2 ceiling becomes trivial.

Even the goal of achieving a ceiling of 450 ppmv becomes vastly more accessible. Intertemporal shifting of emissions alone keeps down costs in the first quarter century, but by the middle of the next century, almost three percent of GDP in China and Eastern Europe and the Former Soviet Union (EEFSU) is consumed by emissions reduction costs. The development and deployment of advanced energy technology by the year 2020 radically reduces these costs to approximately half of one percent of GDP. The assumed continued technological advances eventually reduce these costs to trivial levels. The main economic costs are experienced in China and the EEFSU, and these costs occur during the technology transition period: 2005 to 2050.

6. Technology and protocols to stabilize the atmosphere

The economic penalty associated with following an emissions control trajectory depends on the energy system technologies that are available, and their associated costs. The accelerated introduction of advanced energy technologies can so substantially reduce the costs of meeting an atmospheric CO_2 concentration that costs are insignificant until late in the century. Even the concentration ceiling of 450 ppmv can be achieved at a fraction of the costs required with reference case energy technology. A clear implication of this result is that measures which accelerate global technology cost reduction, development, and deployment have substantial value in achieving the objectives of the FCCC.

Precisely what measures and institutions will realize the above assumptions has not yet been established. Researchers such as Hourcade and Chapuis (1994) and Grubb (1995) have argued that financial signals will lead to induced technological change. The degree to which prices can overcome market failure in the area of research and development (R&D) remains an open question. Arrow (1962) and others such as Levin et al. (1984) and Cohen and Noll (1991) have shown that the inappropriability of economic benefits of R&D leads to chronic underinvestment in technology development. There is thus an important potential role for government intervention. The success of the United States "Golden Carrot" program to develop an energy efficient refrigerator capable of providing the same services as a standard refrigerator at a similar cost indicates the value of a potential role for joint government-private-sector projects. Hourcade and Chapuis (1994) have shown the potential cost-reducing benefits from price-induced technological change. However, Hogan and Jorgenson (1990) have presented evidence for the United States that while induced technological change in the energy sector works at the micro-scale, at the macro-scale feedback effects may actually work to increase costs.

Part of the diversity of the research findings cited above may stem from the fact that behavior differs from community to community in human society. Because institutions and social and economic organizations vary greatly from place to place and over time, there is unlikely to be any single policy implementation strategy which works universally. We leave for others the problem of identifying specific institutional arrangements to accelerate the development and deployment of advanced energy technologies such as those examined here, but note that this is an important problem to address.

The problem of introducing and implementing an emissions ceiling agreement has been studied. Researchers have explored the principles that might guide the development of an emissions stabilization protocol and the broad potential terms and conditions that might be included in agreements (Ghosh, 1991; Grübler and Fujii, 1991; Barrett, 1990; Morrisette and Plantinga, 1990; Morrisette et al., 1991; Sebenius, 1990; Grubb, 1989). Researchers such as Edmonds et al.

(1995b), Grubb (1989). Anderson (1990), Epstein and Gupta (1990), Rose and Stevens (1992), Solomon and Ahuja (1991), and Yamaji (1992) have also examined some of the economic and technical implications of implementing alternative principles.

The work of Edmonds et al. (1995b) indicates that if the costs of emissions reductions are high, then the transfer payment necessary to compensate developing nations for participation in protocols to stabilize emissions become large and potentially daunting to developed nations. But without the active participation of developing nations and economies in transition, any agreement aimed at reducing global emissions or stabilizing the concentrations of greenhouse gases is intractable. Atmospheric CO_2 stabilization is therefore less likely in the presence of high costs of emissions reductions. But high costs can be avoided only if nonemitting technologies can be developed and disseminated globally.

Since both fossil and non-fossil energy technologies are expected to improve under normal conditions, development and deployment require a systematic bias in the development of technologies which do not vent carbon dioxide freely to the atmosphere. From the policy perspective, a protocol must be credible in engendering an expectation that future energy systems will truly eschew free venting of fossil carbon.

7. Conclusions

Implementation of Article 2 of the Framework Convention on Climate Change (FCCC) should be considered explicitly. The FCCC sets as its goal "stabilization of greenhouse gas concentrations in the atmosphere at a level that would prevent dangerous anthropogenic interference with the climate system." While the achievement of this goal is generally considered to be more difficult than achieving the goal of emissions stabilization, this need not be the case. We consider the goal of stabilizing the atmospheric concentration of CO_2 at 450 ppmv and 550 ppmv. We note that stabilization of *emissions* at current levels leads to achievement of this concentration shortly after the turn of the next century (year 2100), though concentrations continue to rise for an extended period of time thereafter. By comparison, we note that the present discounted costs of stabilizing concentrations at 550 ppmv are lower when undertaken in a program of emissions controls which follows that of Wigley et al. (1996).

The trajectory provides for an initial increase in fossil fuel carbon emissions which closely follows the reference IS92a trajectory until the year 2020. Thereafter the rate of growth of emissions declines sharply. Emissions begin a continuous decline by the year 2050. Lower present discounted costs are possible for this case in contrast to the emissions stabilization case, due to four factors:

1. *Carbon cycle.* Natural sinks have a longer period over which to remove emissions early in the period of analysis. This implies higher cumulative emissions for trajectories which stabilize at a similar concentration but with larger near-term loading. The carbon cycle dividend can be substantial. For the 550 ppmv computation the difference between cumulative emissions for the stable emissions and atmospheric stabilization cases is 60 PgC (10 years of fossil fuel carbon emissions at present rates).

2. *Discounting.* The further in the future a given economic burden lies, the smaller the resources that must be set aside today to facilitate the activity. There is great leverage to be gained by postponing the burden of emissions reductions. This is not the same as procrastinating. Resources still need to be put in place to accommodate the future burden, but those resource requirements are lessened.

3. *Technology development.* Emissions reduction technologies are not static. As the IPCC Working Group II notes, multiple technologies exist in various stages of development and deployment which have the capacity to reduce emissions in the future at decreasing costs. By timing emissions mitigation to coincide with technology development, overall costs of stabilizing the concentration of CO_2 can be reduced.

4. *Capital utilization.* Both physical and human capital are created with a set of expectations about the environment in which they will be used. These expectations shape the character of these investments. Existing investments were made with no expectation of a carbon emission penalty. There is a cost to utilizing capital in an environment for which it was not configured. As a general rule, the greater the departure from the anticipated operating environment, the larger the penalty sustained. For atmospheric stabilization ceilings which do not require immediate emissions reductions, there is a benefit to announcing future emissions controls, but implementing them over time.

We find that the character of technologies that are available and capable of global dissemination at the end of the next quarter century (year 2020) have a major impact on the costs of stabilizing the atmosphere. The widespread deployment of technologies which mirror the cost and performance characteristics of those described in the IPCC Working Group II Second Assessment Report promises to enable the concentration of CO_2 in the atmosphere to be held to 600 ppmv without further intervention. Stabilization at 550 ppmv requires no intervention at all until after the year 2050, and then the economic burden is low, less than 0.2 percent of GDP.

In considering the implications for agreements to control the atmosphere, we find that framing the problem as a staged activity, with the principal features of Stage I being technology development and deployment, Stage II being emissions stabilization, and Stage III being phase-out of carbon-emitting technologies, holds some promise. This strategy is potentially useful in implementing ceilings of 500 ppmv or higher. Critical to the success of such a strategy is Stage I.

If nonemitting technologies can be developed and deployed on a global scale, at costs which are comparable to present technology costs, then the problem of implementing Stages II and III are minimal. If the cost structure of energy systems remains as it is assumed to be in the reference case IS92a, then substantial emissions control costs will be encountered, and these costs could imply large burdens and wealth transfers to implement an agreement. It is important to note, however, that large-scale development and deployment of energy technologies which do not freely vent carbon to the atmosphere is unlikely without an agreement to limit fossil carbon discharge which is credible to planners.

Disclaimer
While we are grateful for the financial support for the conduct of our research provided by the Electric Power Research Institute (EPRI) and the United States Department of Energy (DOE), the views expressed in this paper are solely those of the authors and were formed and expressed without reference to positions taken by these institutions. The views of the authors are not intended either to reflect or imply positions of either EPRI or DOE.

Acknowledgments
We would like to express our appreciation to William Chandler and Howard Gruenspecht with whom we have shared early versions of these ideas and whose insights have helped improve the direction and content of this paper. We would also like to express our appreciation to the Electric Power Research Institute (EPRI) and to the United States Department of Energy (DOE). This work was supported in part by the EPRI under contract number DE-AC06-76RLO1831 and the DOE under contract number DE-AC06-76RLO1830.

References

Anderson, D (1990) *International Aid and the Environment: With Special Reference to the Global Warming Problem.* London: University College and Oxford: St Antony's College.
Arrow, K (1962) Economic welfare and the allocation of resources to invention. *In:* NBER: *The Rate and Direction of Inventive Activity: Economic and Social Factors.* National Bureau of Economic Research, Princeton, NJ, 609–626.
Barrett, S (1990) *The paradox of international agreements.* Presented to the American Economic Association Meeting, 29 December 1990, Washington, DC.
Cohen, LR and Noll, RG (1991) *The technology pork barrel.* The Brookings Institution, Washington, DC.
COP (The Conference of the Parties) (1995) The Berlin Mandate: Decision 1/CP.1. reprinted in *The United Nations Climate Change Bulletin,* Issue 2, 2nd Quarter 1995. Interim Secretariat of the UNEP/WMO Intergovernmental Panel on Climate Change (IPCC), and the UNEP/WMO Information Unit on Climate Change (IUCC), Geneva Executive Center, CP 356, 1219 Chatelaine.
Edmonds, J and Reilly, J (1985) *Global Energy: Assessing the Future,.* Oxford University Press, New York.
Edmonds, J, Wise, M, and MacCracken, C (1994) *Advanced Energy Technologies and Climate Change: An Analysis Using the Global Change Assessment Model (GCAM).* PNL-9798, UC-402. Pacific Northwest Laboratory, Richland.
Edmonds, J, Wise M, Sands R, MacCracken C, and Pitcher H (1995a) *Agriculture, Land-Use, and Energy: An Integrated Analysis of the Potential Role of Biomass Energy for Reducing Potential Future Greenhouse Related Emissions.* Pacific Northwest Laboratory, Washington DC.
Edmonds, J, Wise, M and Barns, D (1995b) Carbon coalitions: the cost and effectiveness of energy agreements to alter trajectories of atmospheric carbon dioxide emissions. *Energy Policy* 23(4/5):309–336.
Epstein, JM and Gupta, R (1990) *Controlling the Greenhouse Effect: Five Global Regimes Compared.* Brookings Occasional Papers. The Brookings Institution, Washington, DC.
Ghosh, P (1991) *Structuring the Equity Issue in Climate Change.* Tata Energy Research Institute, New Delhi.
Grubb, M (1989) *The Greenhouse Effect: Negotiating Targets.* The Royal Institute of International Affairs, London, Energy and Environment Programme.

Grubb, M (1992) *The Costs of Climate Change: Critical Elements.* Paper presented to the International Work-shop on Costs, Impacts, and Possible Benefits of CO2 Mitigation, 28–30 September 1992, IIASA, Laxenburg, Austria.

Grubb, M (1995) What You Don't Know Can Hurt You: Scale and Timing of Options in Responding to Climate Change. Paper presented to the IPECA Symposium on Climate Change: A Petroleum Industry Perspective 12–15 April, 1995, Rome.

Grübler, A and Fujii, Y (1991) Inter-generational and spatial equity issues of carbon accounts. *Energy* 16(11/12):1397–1416.

Hourcade, J-C and Chapuis, T (1994) No-regret potentials and technical innovation: a viability approach to inte-grated assessment of climate policies. *In*: N Nakičenovič, WD Nordhaus, R Richels and FL Toth (eds): *Integrative Assessment of Mitigation, Impacts, and Adaption to Climate Change.* CP-94-9. IIASA, Laxenburg.

IPCC (Intergovernmental Panel on Climate Change) (1990) *Scientific Assessment of Climate Change.* World Meteorological Organization/United Nations Environmental Program. Cambridge University Press, New York.

IPCC (Intergovernmental Panel on Climate Change) (1992) *Climate Change 1992: The Supplementary Report to the IPCC Scientific Assessment.* Cambridge University Press, Cambridge.

IPCC (Intergovernmental Panel on Climate Change) (1995) *Climate Change 1994: Radiative Forcing of Climate Change and An Evaluation of the IPCC IS92 Emissions Scenarios.* Cambridge University Press, Cambridge.

IPCC (Intergovernmental Panel on Climate Change) (1996) *Climate Change 1995: Impacts, Adaptation, and Mitigation of Climate Change: Scientific-Technical Analysis. The Contribution of Working Group II to the Second Assessment Report of the Intergovernmental Panel on Climate Change.* Cambridge University Press, Cambridge.

Kosobud, RF, Daly, TA, South, D, and Quinn, K (1994) Tradable cumulative CO2 permits and global warming control. *Energy J.* 15:213–232.

Leggett, J, Pepper, WJ, Swart, RJ, Edmonds, J, Meira Filho, LG, Mintzer, I, Wang, MX, and Wasson, J (1992) Emissions Scenarios for the IPCC: An Update. *In*: IPCC: *Climate Change 1992: The Supplementary Report to the IPCC Scientific Assessment.* Cambridge University Press, Cambridge.

Levin, RC, Cohen, WM, and Mowery, DC (1984) RandD appropriability, opportunity, and market structure: New evidence on some schumpeterian hypotheses. *Am. Econ. Rev.* 75:20–24.

Manne, AS and Richels, R (1995) *The Greenhouse Debate – Economic Efficiency, Burden Sharing and Hedging Strategies.* Electric Power Research Institute, Palo Alto.

Morrisette, PM and Plantinga, AJ (1990) *How the CO_2 Issue is Viewed in Different Countries.* Discussion Paper ENR91-02. Resources for the Future, Washington, DC.

Morrisette, PM, Darmstadter, J, Plantinga, AJ, and Toman, MA (1991) Prospects for a global greenhouse gas accord: Lessons from other agreements. *Global Environmental Change* 1: 209–223.

Nordhaus, WD (1979) *The Efficient Use of Energy Resources.* Yale University Press, New Haven.

Richels, R and Edmonds, J (1994) The Economics of Stabilizing Atmospheric CO_2 Concentrations. *Proceedings of the Tsukuba Workshop of IPCC Working Group III*, CGER-IO12-94, Tsukuba, JAPAN, 17–20 June.

Rose, A and Stevens, B (1992) *The Efficiency and Equity of Marketable Permits for CO_2 Emissions.* Department of Mineral Economics, The Pennsylvania State University.

Sebenius, JK (1990) *Global Environmental Policy Project: Negotiating a Regime to Control Global Warming.* John F. Kennedy School of Government, Harvard University, Cambridge.

Solomon, BD, and Ahuja, DR (1991) International reductions of greenhouse-gas emissions: An equitable and effi-cient approach. *Global Environ. Change* 1(5):251–364.

Toronto Climate Conference Statement (1988) *The Changing Atmosphere: Implications for Global Security.* Toronto, June 27–30, 1988.

United Nations (1992) *Framework Convention on Climate Change.* United Nations, New York.

Wigley, TML, Richels, R, and Edmonds, JA (1996) Economic and environmental choices in the stabilization of atmospheric CO_2 concentrations. *Nature* 379(6562):240–243.

Yamaji, K (1992) *A Simulation Study on International Trade of CO_2 Emission Permits.* Presentation to the IIASA International Workshop on Energy/Ecology/Climate Modeling and Projections, Laxenburg.

Cost-Benefit Analyses of Climate Change: The Broader Perspective
F.L. Toth (ed.)
© 1998 Birkhäuser Verlag Basel/Switzerland

Policy context: Follow-up of the Berlin climate conference

Michael Ernst

Federal Ministry for the Environment, Nature Conservation and Nuclear Safety, Bernkasteler Str. 8., D-53175 Bonn, Germany

1. Introduction

Global climate protection is one of the most important challenges for environmental policy which confronts policymakers, industry and society alike. Climate protection can only be effective if the international community cooperates. One country alone will not be able to ward off the greenhouse effect. The United Nations Framework Convention on Climate Change, which entered into force on 21 March 1994 and which creates an internationally binding basis for global climate protection, was an important step towards such an internationally concerted approach. About 150 countries have now ratified the Convention.

In Article 2 of the Convention, the objective of stabilizing the greenhouse gas concentrations in the atmosphere at a level that would prevent dangerous anthropogenic interference with the climate system is laid down. To achieve this objective, all Parties have general commitments according to Article 4.1. Developed country Parties included in Annex I of the Convention are obliged to return to their 1990 greenhouse gas emission levels by the year 2000 (Art. 4.2 a and b); the Convention does not contain any specification as to what is to happen beyond 2000. The developed country Parties included in Annex II of the Convention also have financial commitments (Art. 4.3–4.5) to assist developing country parties in fulfilling their obligations under the Convention. In Article 11, a financial mechanism is defined; the Global Environmental Facility has been entrusted on a preliminary basis with the operation of the financial mechanism. According to Article 12, developed and developing country Parties have different reporting commitments. In fulfilling their obligations, the Parties shall be guided by the principles laid down in Article 3, e.g. the precautionary approach in Article 3.3.

2. The Berlin Conference

The First Session of the Conference of the Parties to the United Nations Framework Convention on Climate Change took place in Berlin from 28 March to 7 April 1995. This conference was central to the follow-up process of the Framework Convention on Climate Change. As a key element, the Conference dealt with reviewing the commitments of the developed country Parties pursuant to Article 4.2 a and b of the Convention.

The Berlin Conference of the Parties concluded that the commitments of the developed country Parties are not adequate and have to be strengthened in order to counter the global greenhouse effect effectively. After very difficult negotiations, it was possible to adopt a mandate with guidelines and objectives for the negotiation of a protocol on another binding legal instrument.

The "Berlin Mandate" is a first step towards strengthening the commitments pursuant to the Convention and contains the following key elements:

1. Policies and measures as well as qualified limitation and reduction objectives for greenhouse gas emissions within specified time frames such as 2005, 2010 and 2020, are to be laid down for the developed country Parties. Thus the Berlin mandate takes account of the particular responsibility of the developed countries.
2. For the developing country Parties, new commitments are not to be introduced at the moment. At the same time, however, the existing general commitments in Article 4.1, which also apply to the developing country Parties, are reaffirmed and their implementation is to be advanced.
3. A newly established working group has been entrusted with conducting the negotiations, the Ad-hoc Group on the Berlin mandate (AGBM).
4. The draft protocol presented by the Alliance of Small Island States (AOSIS), along with other proposals and already available documents, should be included in the negotiations.
5. The protocol is to be adopted during the Third Session of the Conference of the Parties to the Convention in 1997.

Another important result of the First Session of the Conference of the Parties was that an agreement on a pilot phase of "activities implemented jointly" could be achieved. This means that greenhouse gas emissions could be reduced in a more cost-efficient way through cooperation with the Central and Eastern European or developing countries than through taking measures in a developed country. This pilot phase is open for voluntary participation to all Parties to the Convention taking an interest in it. During this phase no Party will be credited with emission reductions achieved through pilot projects in other countries. On the basis of the experience gained during the pilot phase, the Conference of the Parties to the Convention is to decide by the end of the decade on how the concept "activities implemented jointly" is to be followed up after the conclusion of the pilot phase.

The AGBM started its work in August 1995 in an organizational session. The first negotiation round which dealt with substantial matters was held in Geneva from 30 October to 3 November 1995.

The first two sessions of the AGBM have already shown that the further negotiations on how to give the protocol a specific shape will be extraordinarily difficult given the wide differences in the positions taken by countries. The form which the limitation and reduction targets are to take as well as the definition of specific policies and measures, will be the subject of particularly controversial discussions.

The proposal for a protocol structure presented by the European Union at the second session of the AGBM was given much attention and met with a positive reaction, above all from the part of the developing countries, including the Alliance of Small Island States. This proposal is based on an Anglo-German draft and provides for commitments of the developed country Parties regarding quantified limitation and reduction objectives, as well as policies and measures as elements for the limitation and reduction of greenhouse gas emissions from all relevant sectors and for the protection and enhancement of greenhouse gas sinks. The commitment regarding policies and measures should be specified through three annexes. The first would include policies and measures which all Annex I Parties agreed to adopt and include in their national programs. The second would contain a list of policies and measures which it was agreed would benefit from early international coordination, and which should be given high priority consideration for inclusion in national programs. The third would include policies and measures which are identified as having proven effectiveness or potential, and which should be considered for inclusion in national programs, as appropriate to national circumstances.

In addition, the proposal for a protocol structure provides for continuous reporting on the implementation of the protocol obligations as well as its regular review and further development.

Evidence presented by scientists and increasingly being confirmed compels us to take consistent precautionary action. The Intergovernmental Panel on Climate Change (IPCC) adopted its Second Assessment Report in December 1995 in Rome. This report contains the state-of-the-art with respect to the science of climate change, the impacts of climate change as well as adaptation and mitigation options, and the economic and social dimensions of climate change. An important finding of this report is that the balance of evidence suggests a discernible human influence on global climate. Furthermore, if no counter measures are taken the mean global air temperature is projected to increase by 1 to 3.5 degrees Celsius by the year 2100 as compared to 1990. The sea level is projected to rise by 15 to 95 cm on average. The Second Assessment Report also contains a synthesis of scientific-technical information relevant to interpreting Article 2 of the Convention. This synthesis can be seen as a first step in approaching Article 2 from a scientific-technical view. A detailed discussion of Article 2 will probably not begin before finalizing the work of the AGBM.

As far as the Convention is concerned, there are three groups of countries:

- Annex I countries (OECD countries, the European Union and countries with economies in transition, i.e. countries of the Former Soviet Union and Eastern Europe)
- Annex II countries (Annex I countries minus countries with economies in transition)
- non-Annex I countries (developing countries)

In the negotiations during sessions of the Convention bodies, a number of groups of countries conduct meetings, whereby the positions are bundled to reach decisions:

- European Union (EU)
- JUSCANZ (Japan, US, Canada, New Zealand and other non-EU Annex II countries)
- Common Interest Group (EU members and JUSCANZ countries)
- Group of 77 (G77, group of about 130 developing countries)
- Alliance of Small Island States (AOSIS)
- Oil exporting developing countries (OPEC countries)

The formation of these groups is flexible, for example a so-called Green Group which tabled a proposal for the later Berlin Mandate at the Berlin climate conference, initially consisting of 42 developing countries belonging to the G77, ultimately contained more than 70.

In plenary meetings of the Convention bodies, the countries and the Presidency of the EU as well as the Chair of G77 are speaking on behalf of their groups. To come to agreements in a session, it is often useful to meet with a number of key countries from each of the above-mentioned groups; this has been done in the case of negotiating the Berlin Mandate, holding meetings with 24 countries. Besides these groups, the five UN regional groups (Africa, Asia and Pacific, Eastern Europe, Latin America and Caribbean, and Western Europe and Others) conduct meetings on personal issues rather than on matters of substance.

3. Forthcoming events

The third and fourth sessions of the AGBM will take place in March 1996, immediately before the Second Session of the Conference of the Parties in Geneva in July 1996. At the third session, *inter alia*, the Second Assessment Report of IPCC will be discussed, concerning policies and measures and emission objectives. Two informal workshops on these issues will also take place. The AGBM also requested the secretariat of the Convention to prepare, for consideration at the Fourth Session, a document reviewing possible indicators that could be used to define criteria for differentiation among Annex I countries with respect to emission objectives. Another important input to the work of the AGBM comes from the "Annex I Expert Group" which is supported by the secretariats of OECD and IEA; it is conducting a so-called common actions study on policies and measures.

To achieve ambitious protocol commitments, the prospect of implementing the current commitments pursuant to the Convention (especially the commitment for developed country Parties to return to their 1990 greenhouse gas emission levels by the year 2000) is crucial; this will be made obvious through compilation and synthesis reports of the national reports of Annex I country Parties.

Together with its partners in the European Union, Germany is making every effort to ensure that ambitious protocol obligations are reached at the Third Session of the Conference of the Parties in 1997. The national climate protection program of Germany already goes much further than is required by the obligations of the Convention. In his speech held during the Berlin climate conference, the Federal Chancellor Dr Helmut Kohl confirmed that Germany continues to work towards its target to reduce its 1990 level of CO_2 emissions by 25% by the year 2005.

The implications of including sulfate aerosols on scenarios of admissible greenhouse gas emissions[*]

Christian Issig[1], Hans-Jochen Luhmann[2] and Paul Vradelis[3]

[1]Meteorological Institute of the University Bonn, Auf dem Hügel 20, D-53121 Bonn, Germany
[2]Wuppertal Institute for Climate, Environment, Energy GmbH, Döppersberg 19, D-42103 Wuppertal, Germany
[3]Institute for Environmental Physics, Heidelberg University, Im Neuenheimer Feld 366, D-69120 Heidelberg, Germany

1. Introduction

The tolerable window approach reported by the German Advisory Council on Global Change (WBGU, 1995) is a useful representation to derive future emission paths compatible with the aim of the FCCC. However, since it involves a climate model that neglects negative radiative forcing from sulphate aerosols in the troposphere, which partially offset the warming induced by greenhouse gases (GHGs), it tends to produce a margin of safety that may be too small. The model does not allow for the fact that, since fossil fuel burning will be reduced as a result of climate protection policy, emissions of sulphur will decrease as well, reducing the cooling effect of these aerosols. As a consequence, the scope of emission paths in the window's interior is narrower than predicted by the simple model of the WBGU.

This consideration can formally be included into the tolerable window approach using the simple WBGU climate model by defining a *virtual point,* representing the present state of the climate system. It shows what the temperature and its rate of change would be at present if fossil fuel-related aerosol cooling were to be switched off, as a policy geared towards reducing fossil fuel burning would imply. Here the contrary effect of a successful climate protecting policy is anticipated. In finding optimal emission paths by model calculations, we should start from this virtual point. The consequences of such considerations for admissible emission scenarios are shown qualitatively. They should be studied more carefully by explicit calculations with suitable models.

---

[*] Editor's note: The tolerable climate window adopted by the WBGU for its initial analysis to derive emission scenarios was less restrictive than the one considered in this paper. In addition to the 16.6 °C upper limit for global mean temperature, WBGU specified 0.2 °C/decade as the maximum rate of temperature increase while Issig et al. take 0.1 °C/decade as their admissible value (cf. WBGU, 1995: 115). In fact, the smaller climate window discussed in this paper was also examined by the WBGU in a sensitivity analysis. However, the WBGU explicitly "points out that a very pessimistic assessment of the adaptability of ecosystems was taken as the basis for this reduced tolerable climate window" (WBGU, 1995: 118).

2. Climate window – base case

In order to stabilise concentrations of greenhouse gases in the atmosphere so as to prevent dangerous anthropogenic interference with the climate system, a target level has to be set up within a time-frame sufficient to allow ecosystems to adapt naturally to climate change. That is the central obligation established by Article 2 of FCCC.

Crucial aspects of the "climate system" mentioned in Article 2 of FCCC can roughly be described by two factors, the global mean temperature (°C) and the rate of its change (e.g. °C/decade). As a necessary condition to ensure that ecosystems will be able to adapt naturally, these two factors must not exceed some threshold.

These constraints can be visualised by the tolerable window approach, as discussed in WBGU (1995:111). The first factor describing the climate system, the global mean temperature, should not exceed a value higher than the highest mean temperature in the current geological period (Quaternary), because this period has moulded the ecological system we are living in. The highest value of about 16.1 °C prevailed in the Eem interglacial period. Admitting a range of tolerance of about 0.5 °C, a maximum admissible global mean temperature would be 16.6 °C. With respect to the second factor, the rate of temperature increase, the Enquete Commission "Protecting the Earth's Atmosphere" of the German Parliament has put forward a maximum admissible value of 0.1 °C per decade (Enquete-Kommission, 1995:96). This is assumed to be the extreme value which allows ecosystems and largescale vegetation zones to migrate in step with a tolerable rate of temperature change. However, as temperature increases, the ability of ecosystems to adapt naturally tends to decrease. At higher temperatures, the tolerable rate of variation in temperature tends towards zero. This feature is also assumed in WBGU (1995:112).

This relationship can be illustrated by means of a temperature window, whose boundaries describe the limiting values and whose inner domain represents the tolerable states of the climate system, defined here by the mean temperature and the rate of its change.

The graph inside the temperature window shown in Figure 1 depicts the increase of global temperature during the last 130 years, together with its rate of change (data from Hadley Centre, 1995). Four phases can easily be distinguished which characterise the observed temperature trend since 1860. In phases I (1860–1910) and III (1940–1970) no significant global warming of the earth's atmosphere occurred. Phases II (1910–1940) and IV (1970–1990) are characterised by an increase of global mean temperature by an average of 0.06 °C per decade. At present the global mean temperature is 15.3 °C and the rate of temperature change is approximately 0.07 °C per decade. The concentration of greenhouse gases in the atmosphere, measured in CO_2 equivalents, must not exceed a threshold value in order to allow the climate system to remain inside the boundaries given by the temperature window.

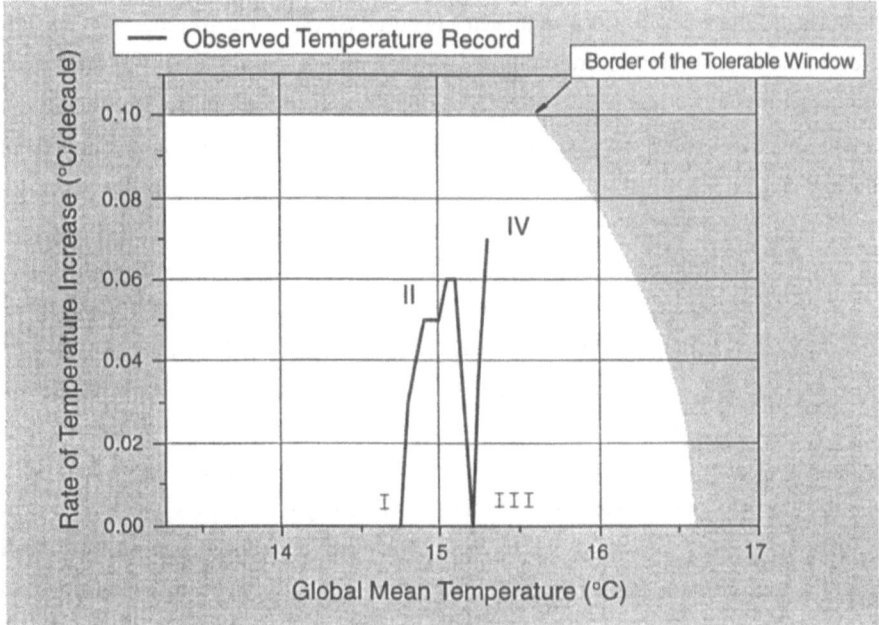

Figure 1. Temperature window and observed global temperature over the last 130 years.

The analysis in WBGU (1995) continues by choosing an emission scenario, examining its future implications on the climate system by means of a calculation using a simple global climate model (GCM), and adapts the emission scenario iteratively until a scenario is attained in which the climate system (among other things) remains within the tolerable window.

However, since the climate model employed by WBGU (1995) explicitly neglects external forcing other than the course of anthropogenic CO_2 emissions, its predictions can be improved by including the effect of aerosol loading of the atmosphere as a consequence of fossil fuel burning. This extension is reasonable since the neglect of the connection between CO_2 and aerosol emissions is misleading, as will be shown in the next section.

3. Climate window including aerosols

When comparing observed global temperature trends with GCM simulations, which only include the positive radiative forcing of GHGs (i.e. neglect the negative forcing associated with aerosols), the discrepancy is apparent: typically they give estimates of the global mean temperature that are too high. These differences between simulated and observed values can be reduced by including the effect of the sulphate aerosols from sulphur dioxide emissions, which are a typical by-product of fossil fuel burning (see also IPCC, 1996:112). This inclusion results in a temperature estimate closer to actual observations. Figure 2 gives examples of the results of both kinds of simulations, i.e. considering

(1) the effect of greenhouse gases only and

(2) the effect of both greenhouse gases and the sulphate aerosols.

Obviously there is a greater similarity of shape between curve (2) in Figure 2 and the curve in Figure 1, than between curve (1) in Figure 2 and the "real" curve in Figure 1.

Obviously, the actual observed temperature trend is modelled more realistically when aerosol forcing is taken into account. Similar results are quoted in IPCC (1996:320). Furthermore, future temperature projections are generally lower when aerosol forcing is taken into account.

This implies that the model employed in the WBGU study (1995) does not provide a large enough margin of safety, since it assumes a decrease in the GHG-induced warming (i.e. positive forcing), but no change in aerosol cooling (i.e. negative forcing). This is not realistic, since a substantial fraction of the sulphur dioxide emissions which are the cause of aerosol loading is a

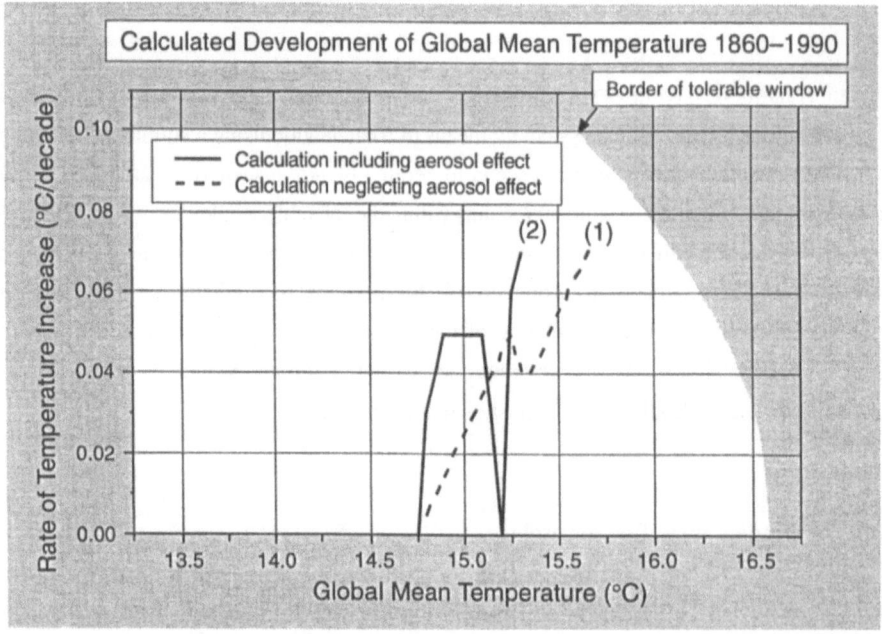

Figure 2. Temperature changes modeled with and without sulphate aerosols.

consequence of fossil fuel burning. Therefore, one is led to expect that temperature increase will be stronger than predicted by the WBGU calculations, which means that the scenario recommended by the WBGU, which is modelled to take the climate system just to the limits of the tolerable window, will actually result in the climate system leaving the window. Put another way, the fact that the cooling aerosol forcing will be reduced along with decreasing fossil fuel burning implies that emissions must be still lower than suggested by WBGU (1995) in order to keep the climate system within the tolerable window.

An obvious solution to this problem would be to employ a climate model which explicitly takes the aerosol effect into account. This can be done as, for example, reported by IPCC (1996:320). However, these analyses merely project the climatic effect of different *hypothetical* emission profiles. In contrast, the tolerable window approach contains explicit quantitative assumptions on admissible impacts on the climate system and can therefore help to identify a *recommended* emissions path. In this paper, we therefore suggest a method of including the dependence between aerosol and CO_2, which does not require a climate model that explicitly simulates aerosol cooling, yet still arrives at a recommended emissions profile compatible with a movement inside of the tolerable window.

This can be done by assuming that sulphur emissions, and hence the negative forcing associated with the resulting aerosols, will be reduced to the same extent as fossil fuel burning over the course of emission abatement. So looking for a suggestive approximation, the starting point for the recommended future development, which represents the "present situation", should not be the point given by the actually measured data, but a point which represents the present situation minus the known aerosol cooling effect. The justification for this is that, due to their short atmospheric lifetimes (of the order of a few days), decreases in emissions of the aerosol precursors show an almost immediate effect on the atmospheric concentration of the aerosols and therefore on the associated cooling, which is in contrast to the greenhouse gases, which have much longer atmospheric lifetimes (e.g. CO_2: 50–200 years) and whose warming therefore persists for this time span. In this way we arrive at a *virtual point* in the temperature window which then represents a suitable starting point for finding a future emission path compatible with the restrictions imposed by the tolerable window. This virtual point will be shifted from the actually observed point towards the values predicted by models which disregard aerosol cooling (see Fig. 3). Since this point lies closer to the limits of the tolerable window it is obvious that reductions in emissions must be greater than those recommended by calculations starting at the actual point and neglecting aerosol cooling.

As mentioned previously, in order to obtain quantitative results, it would be preferable to employ a model that fully accounts for the aerosol cooling and to assume a realistic relationship between CO_2 and sulphur emissions. However, since this requires complicated GCM calculations,

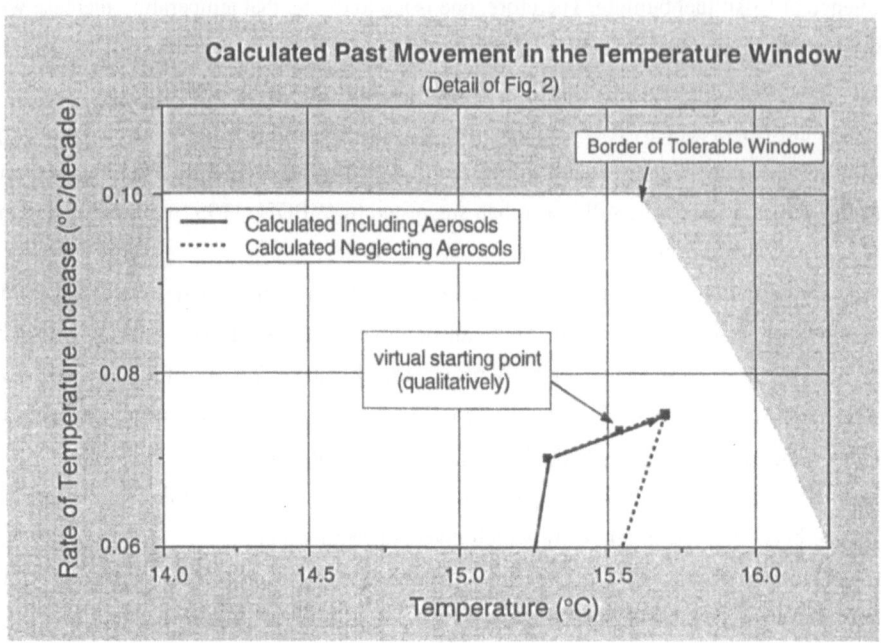

Figure 3. The virtual point in the tolerable window as starting point for future development.

some rough estimates are given here. They are based on quite elementary models and include substantial simplifications and approximations and are therefore to be considered as a guideline only.

To obtain more realistic results we have to make explicit assumptions regarding the linkage
between sulphur emissions and fossil fuel burning. Here, we assume that sulphur emissions will
decrease proportionally to the decline in fossil fuel consumption, i.e. the energy-related CO_2
emissions, although future reductions are expected to be even more severe due to targeted specific
sulphur emission mitigation efforts through fuel switching (from coal to natural gas) and end-of-
the-pipe technologies such as flue gas cleaning. This assumption is in keeping with the spirit of
the analysis in WBGU (1995: 113), which tries to avoid too pessimistic results for climate
protection policy, and aims at a "lower boundary" for the reduction requirements.

The virtual point, which represents the starting point at present, is determined in the following
way. The magnitude of the negative radiative forcing caused by sulphate aerosols is at present
approximately 1.1 Wm^{-2} (IPCC, 1996: 320). The positive GHG forcing is estimated to be
2.75 Wm^{-2} (IPCC, 1996: 21), so that the net effect is a positive 1.65 Wm^{-2}. Simply adding up
positive and negative components of radiative forcing is a coarse approximation, but suffices for
the guidelines we wish to derive here. If fossil fuel-related emissions are eventually cut by some

Interpolated Depedence of Temperature on Radiative

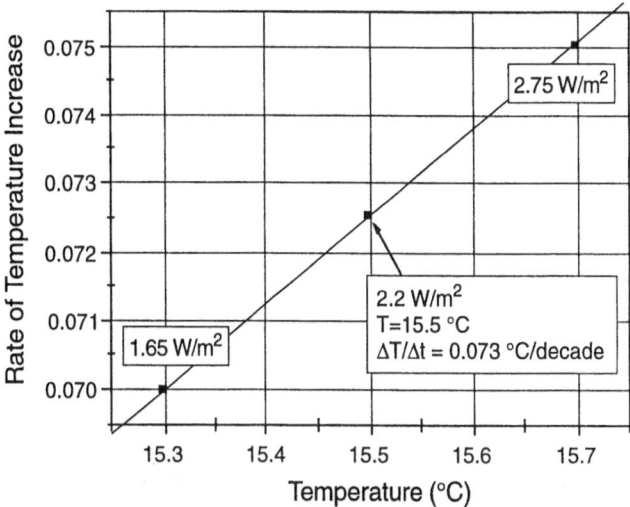

Figure 4. Determining the virtual starting point by interpolation.

50% (as is assumed in the emission path suggested by WBGU) between 1995 and 2005, the negative aerosol forcing will, as a first approximation, be decreased by the same fraction to 0.55 Wm^{-2}. This is assumed to be the aerosol radiative forcing remaining in the final state. If the aerosol emissions were cut down to the above value, the present forcing would be 2.2 Wm^{-2}. The 1990 point in the calculations in Figure 1, which neglect the aerosol forcing, consequently corresponds to a forcing of 2.75 Wm^{-2}, the actual point to 1.65 Wm^{-2}. The virtual point chosen to correspond to the expected decline in aerosol cooling should then be represented by a point which lies on a straight line between these two points, just as far from the actual point as the "virtual" forcing of 2.2 Wm^{-2} lies from the actual forcing of 1.65 Wm^{-2}, if the other end of the line corresponds to the greenhouse gas only forcing of 2.75 Wm^{-2} (see Fig. 4). This approximation yields the virtual point (15.5 °C; 0.0725 °C/decade) as a representation of the starting point at present.

Our next step is to investigate the consequences of this new situation for the recommended emission scenario. For this purpose the recommendation stated in WBGU (1995: 117) is repeated once more (Fig. 5). The development of the temperature given in this figure corresponds to an emission scenario which involves an initial ten year period with an emission reduction of 6.3% per year, followed by a steady reduction of 0.3% per year up to 2195. The virtual starting point is shown in Figure 5 as well. It can easily be seen that reductions must be greater when the temperature development is correctly perceived as starting at this point.

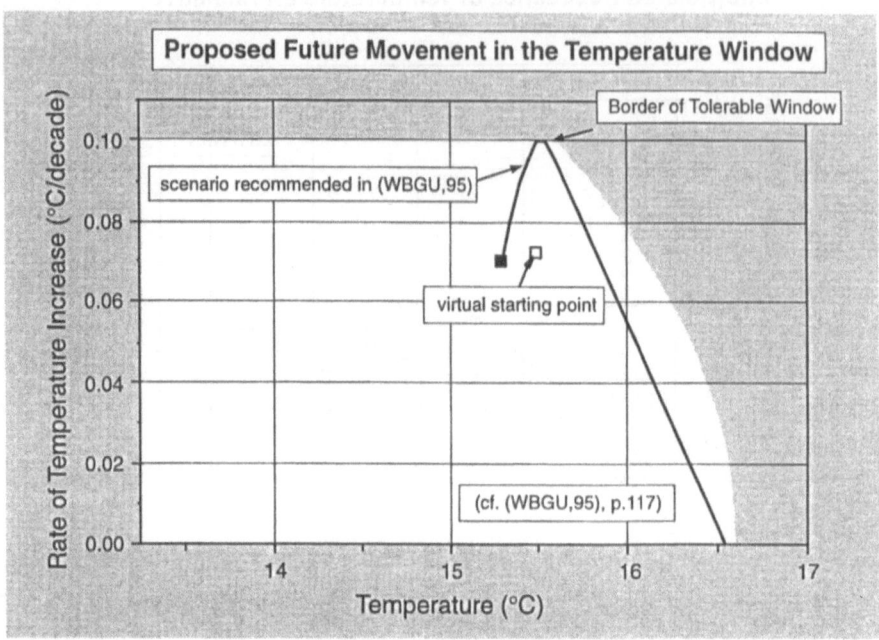

Figure 5. The scenario presented in WBGU (1995) with the new virtual starting point.

To estimate the magnitude of these additionally necessary reductions we assert that the cumu-lated emissions of CO_2 must be chosen so that the increase in radiative forcing by GHGs is lower than in the case of the recommended scenario by the same amount that the aerosol forcing will decrease by the reductions. The relationship between the additional radiative forcing and the cumulated emissions is roughly approximated by a linear dependence, which is a simplification, but in good agreement with the dependence found in IPCC (1996:320, 326) using the IS92a–f scenarios. This dependence is illustrated in Figure 6.

From this linear approximation it can be estimated that the scenario proposed by WBGU, which anticipates cumulated emissions of 682 GtC, yields an additional greenhouse gases radia-tive forcing of 2.35 Wm^{-2}. Taking into consideration the smaller permissible increase in GHG forcing that results from the 0.55 Wm^{-2} decrease in negative aerosol forcing, the tolerable increase in positive forcing must not exceed 1.80 Wm^{-2}. According to Figure 6, this allows for cumulated CO_2 emissions of 542 GtC. These two scenarios are also shown in Figure 6.

These considerations suggest that only a reduction approximately 20% greater than that proposed by WBGU will result in a temperature trend that remains confined within the tolerable window and is therefore compatible with the aims of the FCCC. The recommended emission

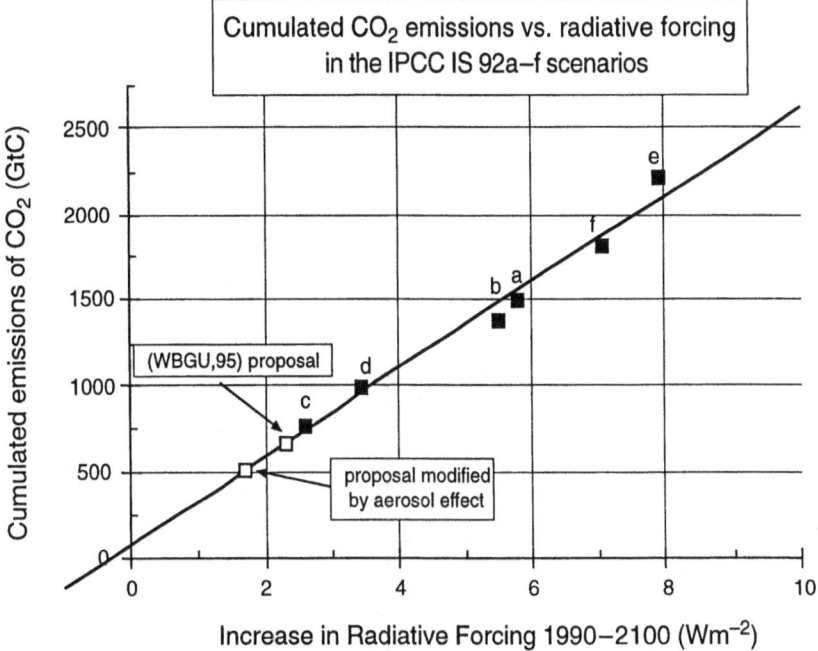

Figure 6. Dependence of admissible cumulated emissions on tolerable increase in radiative forcing.

scenario from the WBGU study has to be adapted accordingly. This is shown in Figure 7. Let us point out once again that these recommendations are based on very simple models and need to be studied more carefully to obtain more accurate results.

4. Conclusions

In summary it may be said that the sulphate aerosols currently offset the global warming by a substantial fraction. Assuming that the aerosol content will decrease in proportion to the emission of GHGs over time as a consequence of declining fossil fuel use, this offsetting effect to global warming will steadily decrease.

Looking for an approximation which can be easily understood, the starting point for the future course, representing the "present situation" in the temperature window, should not be the actual point (15.3 °C, 0.07 °C/decade), which is shown in Figure 1. Rather it should be the virtual point (15.5 °C, 0.073 °C/decade) shown in Figure 5, which stands for the present situation if the "masking effect" of aerosols over time is anticipated. So the situation is more dramatic than the

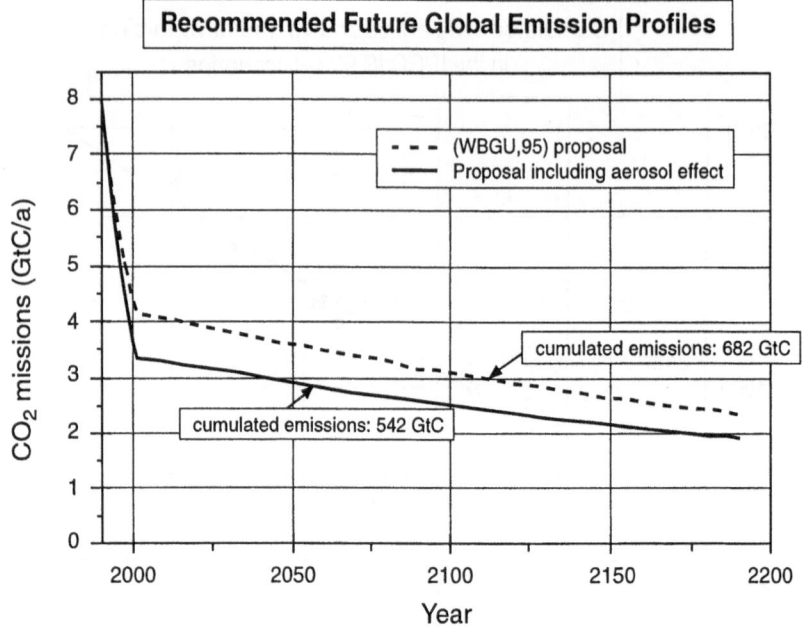

Figure 7. The effect of including the aerosol effect on the recommended emission profile.

observed data show. This implies that the substantial emission reduction requirements suggested by the WBGU have to be ratcheted up even higher.

References

Hadley Centre (1995) *Modelling Climate Change (1860–2050)*. The Hadley Centre, Berks.
Enquete-Kommission (1995) Enquete-Kommission "Schutz der Erdatmosphäre" des Deutschen Bundestages (ed.): *Mehr Zukunft für die Erde. Nachhaltige Energiepolitik für dauerhaften Klimaschutz*. Economica Verlag, Bonn.
IPCC (Intergovernmental Panel on Climate Change) (1996) *Climate Change 1995: Economic and Social Dimensions of Climate Change*. Cambridge University Press, Cambridge.
WBGU (GermanAdvisory Council on Global Change) (1995) *World in Transition: Ways towards Global Environmental Solutions. 1995 Annual Report*. Springer-Verlag, Berlin.

Cost-Benefit Analyses of Climate Change: The Broader Perspective
F.L. Toth (ed.)
© 1998 Birkhäuser Verlag Basel/Switzerland

The tolerable windows approach to climate control: Optimization, risks, and perspectives

Gerhard Petschel-Held and Hans-Joachim Schellnhuber

Potsdam Institute for Climate Impact Research, P.O. Box 601203, D-14412 Potsdam, Germany

1. Introduction

Scientists working in the field of global warming are subject to the increasing tension between soundly based scientific research and the political necessity for the formulation of adequate climate protection strategies. This narrow route between Scylla and Charybdis, which has to prevent science both from withdrawing into its traditional ivory tower and becoming lost in the maelstrom of unbased political claims, demands a well-defined borderline between normative settings and the strict analysis of the Earth System. The borderline is sketched in Article 2 of the Framework Convention on Climate Change (FCCC): "… stabilization of greenhouse gas concentrations in the atmosphere at a level that would prevent dangerous anthropogenic interference with the climate system. Such a level should be achieved within a time frame sufficient to allow *ecosystems to adapt naturally to climate change, to ensure that food production is not threatened, and to enable economic development to proceed in a sustainable manner*" (UN, 1992; emphasis added). It is the task of science to find out how this political constraint can be fulfilled. It is important, however, to keep in mind that – similar to Odysseus – the route outlined by Article 2 is subject to continuous corrections and thus has to be followed by fuzzy control. This is due both to the fact that norms, settings, and society itself are changing quite rapidly, and to the uncertainties in the knowledge of the mechanisms of the system.

Against this background the Berlin Mandate, which was released by the First Conference of the Parties in April 1995, represents an extreme challenge to both the political and the scientific community. The call to the latter which is reflected in the creation of the "Subsidiary Body for Scientific and Technological Advice (SBSTA)" asks for advice on all aspects of anthropogenic climate change: driving forces (identification of emission scenarios), improved understanding of the climate-related aspects of the atmospheric system, and assessments of possible climate impacts. The necessity for full assessments is explicitly mentioned and emphasized as being a possible input by the IPCC (UN, 1995).

Altogether we have to note the high and urgent demand for good integrated assessments of climate change which, according to the first paragraph, have to take care of the explicit borderlines between science and politics. Then these assessments can be used as an information baseline for the difficult attempt to apply fuzzy control to the climate system.

Traditional integrated assessments of global climate policy concentrate on the analysis of potential costs and benefits of such a policy (Kaya et al., 1993; Nakičenovič et al., 1994). They start from the formulation of future scenarios of GHG emissions, i.e. those emission paths which seem to be admissible within the technological, economic and social context (IPCC, 1990). By assessing the costs of a specific scenario compared to a "costless" reference and estimating the benefits of the investigated scenario in terms of climate damages avoided which have to be computed in the same units (in general financial), one can carry out a consistent and integrated evaluation of the specific emission profile (Toth, 1994). By comparison of different scenarios as well as different instruments for reduction (activities implemented jointly, certificates, etc.) (Edmonds et al., 1993) it is possible to give rough estimates of the "best" climate protection strategy.

As well as some advantages, e.g. the formulation of integrated frameworks, this type of analysis of climate policy options has quite a number of shortcomings and difficulties. Besides major methodological problems still to be solved, e.g. modeling the economy, the climate system, or the carbon cycle, which will appear in other approaches as well, there are essential shortcomings which are specific to traditional cost-benefit analysis. These shortcomings include the disregard of important nonmarket values, e.g. health, societal effects or migration. Although it might be claimed that these noneconomic factors cannot be part of economic theories, they have to be taken into account in order to come up with sensible, realistic and acceptable recommendations for policy makers.

The Tolerable Windows Approach proposed here tries to avoid these difficulties by the implementation of tolerance thresholds for general stresses of man and environment. These thresholds – in nature similar to critical loads and based on Article 2 of the FCCC – are then translated into corresponding levels for all the relevant subsystems. The philosophy of the concept is discussed in Section 2. In Sections 3 through 5 we illustrate the approach by an example which makes use of simple models and assumptions.

2. The philosophy of the Tolerable Windows Approach (TWA)

The essential paradigm of the Tolerable Windows Approach is the explicit normative setting of unacceptable impacts, policies and instruments. Therefore this approach is similar to the traditional critical load concept. To begin with, the different thresholds are supposed to meet the requirements set up by Article 2. Yet these requirements are only weakly formulated and further specification of Article 2 as well as other general constraints are necessary. These new constraints might follow the permanently updated knowledge on possible climate impacts, e.g. the latest 1995 IPCC report (IPCC, 1996), as well as new political decisions.

Similar approaches have been used before (Richels and Edmonds, 1994; Wigley et al., 1996; Hope et al., 1993) which, however, reveal more or less systematic shortcomings. The work by Richels, Edmonds, and Wigley is restricted to the specification of unacceptable CO_2 concentrations and therefore takes into account the first part of Article 2, but it does not consider the transition to this – still unspecified – level of GHG concentrations whose consideration is required by the second part of Article 2. Hope et al. started from a time-dependent threshold between temperatures with and without relevant impacts. The specification of the threshold is quite arbitrary, and furthermore the effects of changes in the precipitation patterns and the frequencies of extreme events or the rate of temperature change itself are neglected. This is in contradiction with the fact that we have learned that the global mean temperature is not sufficient for the analysis of possible impacts (Toth, 1994).

In view of the necessary clear-cut interfaces between scientific analysis and political and normative inputs which have been mentioned in the introduction, a more detailed and sophisticated structure of the entire problem of global warming is necessary. The structure which represents the heart of the TWA and which is supposed to meet the different systematic necessities is presented in Figure 1. This concept of an integrated assessment of climate protection strategies is currently pursued in an international project called ICLIPS (Integrated Assessment of Climate Protection Strategies) which is headed by the Potsdam Institute and is funded by the German government. For the sake of simplicity we start with a purely global analysis and present the necessary refinements to a multi-actor system later.

In the first step, the entire system is divided into its essential parts which in Figure 1 are represented by the eight plains. Each plain is just an intimation of a high dimensional function space which contains all possible time paths of the corresponding variables. Below each plain some representatives of these variables are listed. Formally, one element of the space s_j is given by the vector-like function $\bar{x}_j(t) \in \mathfrak{R}^{N_j}$ of time. This division allows us to distinguish clearly between the scientific ingredients of the analysis and the normative settings in the form of thresholds between acceptable and unacceptable developments.

Starting from the left hand side, one has to identify the "tolerable window of climate impacts" in terms of health, basic food supply, water availability, threats to the natural and cultural heritage, and many other factors. It is exactly here that most of the items listed in Article 2 of the FCCC enter. This window allows each impact function $\bar{x}_1(t)$ to be evaluated with respect to its acceptability, i.e. to answer the question of whether the considered time evolution is tolerable or not.

ICLIPS: Inverse Translation Strategy

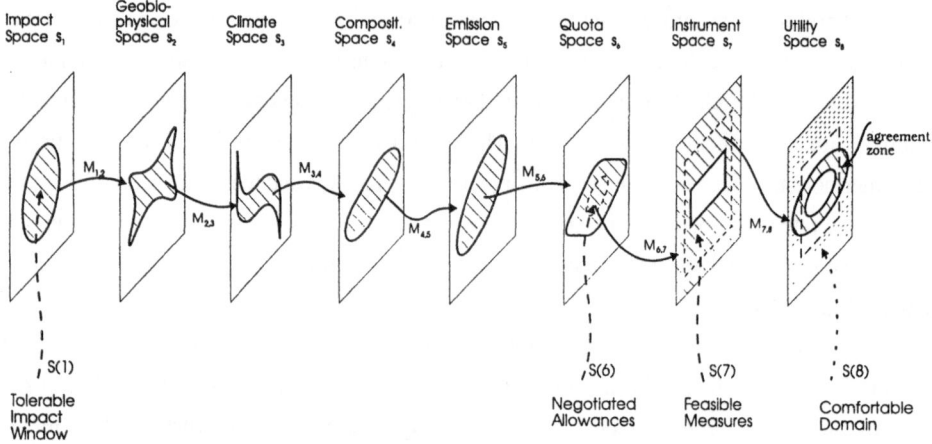

Figure 1. General structure of the Tolerable Windows Approach. Each plain represents one relevant subsystem which can be characterized by the aspects listed below each plain. The normative settings enter as specific tolerable domains and are indicated by the arrow at the bottom. The translations of these domains are represented by the arrows between the plains and are obtained by an inverse mapping strategy. The intersection of the domains in the policy space and in the instrument space results in a general, hopefully non-empty, tolerable window which satisfies all entered norms and values. This final window has to be compared with the window of opportunities derived from the assessment of economic, technological and social options and expectations.

Assuming that the climate impacts are basically a function of biogeophysical parameters, e.g. soil moisture, frequency of storms, solar radiation, temperature, etc., this domain can be "translated" into a corresponding domain of admissible biogeophysical conditions. Formally the new domain is obtained by aid of the operator $M_{1,2}$. This operator generates those functions $\bar{x}_2(t)$ which by the assumptions of causes and effects produce the impact functions $\bar{x}_1(t)$, i.e. $M_{1,2}\,\bar{x}_1(t) = \bar{x}_2(t)$.

In this way, we proceed towards the right hand side of the figure up to the global emissions of greenhouse gases. We have to note that the shape of the admissible domains within each subspace is deeply nontrivial, which is already the case in the simple example given in the next sections. At the level of the analysis of single functions, we end up with a set of emission profiles $\{\bar{x}_5(t)\}$ which correspond to the asserted tolerable impact window. These emission profiles are determined by the successive application of the different operators shown in Figure 1. Formally all these operators are determined by inverting the models for the corresponding subsystems. Therefore we call the entire approach an *inverse mapping strategy*. In this manner, we obtain those emission profiles which meet all the norms and settings in terms of climate impacts.

In the next step the possible quota for the admissible emissions have to be formulated, i.e. which country (or group of countries) should emit how much within what time frame. The corresponding operator $M_{5,6}$ is quite simple: it just has to ensure that the sum over all quotas equals the total emissions. Nevertheless the choice of the quota is not entirely free: according to the FCCC only countries listed in Annex I are committed to reducing their emissions. Therefore the *political setting* determines a domain $\mathcal{D}_{ps}$ of negotiable allowances which has to be contrasted with the domain $\mathcal{D}_6$ of admissible allowances with respect to the climate impacts. In Figure 1 the first domain is sketched as the dark area whereas the latter one is represented by the light grey area. Formally the domain $\mathcal{D}_{quota}$ of acceptable quota with respect to impacts *and* international policies is given as the intersection $\mathcal{D}_{quota} = \mathcal{D}_6 \cap \mathcal{D}_{ps}$.

What are the possible instruments for fulfilling the commitments? What level of taxes do we need, assuming a certain payback mechanism? What type of certificates are proper for the emission targets? Is it sufficient to launch a big education and training program about efficient energy use? This type of question lies at the center of the next step where the domain $\mathcal{D}_7$ of admissible instrument with respect to impacts and policies, has to be determined by use of the operator $M_{6,7}$. But again the domain obtained has to be evaluated with respect to the feasibility of the asserted measure:

- is the necessary level of CO_2 taxes socially acceptable?
- is the required joint implementation and its mechanism politically feasible?

The resulting domain $\mathcal{D}_{FM}$ is sketched as the dark area in the seventh slice in Figure 1 and the intersection yields the domain $\mathcal{D}_{instr}$ of instruments compatible with all entered norms and settings.

Finally the application of the operator $M_{7,8}$ aids the computation of the utilities affected by the instruments obtained in the previous step. These utilities include economic, social, cultural, and political aspects:

- What level of mobility or room heating is admissible?
- What amount of energy services can the economy and its different sectors afford?
- What are the implications for recreation, leisure, and tourism?

This ends up with the life belt-like structure shown in Figure 1 which has to be compared with the comfortable domain, i.e. the desirable level of economic welfare, mobility, energy services, etc. The overlap of these two domains is called the *agreement zone* and contains all the information on admissible climate protection strategies. By propagation of the agreement zone backwards one obtains the corresponding instruments, allowances, emissions, etc. The set of strategies obtained represents scientifically based options for climate protection policies. Ideally, the ongoing negotiations on the Berlin Mandate should pick one of these options.

The discussion so far requires some further comments:

1. Obviously, the procedure as discussed has to been viewed as an ideal one. Knowledge of the different subsystems is still very uncertain and no clearcut formulation of norms and policy options exists. Both these aspects will change over time and therefore the sketched structure is not static but dynamic, i.e. the windows of tolerance as well as the specific mappings between the different spaces change. This results in a subsequent reformulation of the resulting set of strategies and its potential application as fuzzy control.

2. Except for the analysis of different allowances in the quota space, the analysis as discussed so far neglects regional peculiarities. These peculiarities have to be taken into account on the far right and the far left sides of Figure 1. The sensitivity to climate change strongly depends on local conditions and therefore has to be analysed as a regional climate sensitivity. This regional resolution carries through up to the climate subsystem which is mainly governed by global processes. The analysis again decomposes into regions when the different allowances have to be realized by instruments. Similarly the utilities depend on economic, social, cultural, and political aspects and therefore have to be analysed in an appropriate geographic decomposition.

3. The decomposition of the entire system into the suggested eight relevant subsystems is particularly intuitive when the feedback is negligible. As we will see in the simple example in Sections 3 and 4, the inversion of the actual models and thus the formulation of the operators $M_{j,j+1}$, can be achieved analytically. If there is feedback, however, this formulation is not possible. In this case the theory of differential inclusions can be applied (Petschel-Held et al., 1997; Toth et al., 1997)

4. The approach actually opens a set of possible modeling options: so far the operators $M_{j,j+1}$ were designed to translate the corresponding functions directly between the different subspaces. It is, however, possible and might even be more straightforward and feasible to formulate operators which merely translate the domains into each other. The advantage would be that any uncertainties involved are relevant at the border of the domain only.

As was mentioned before, the approach sketched in this section is currently pursued in the ICLIPS project. Nevertheless the basic ideas can be illustrated in an example which makes use of simple assumptions and models. This example is treated in the rest of the article.

3. A simple example: The WBGU scenario

We want to illustrate the Tolerable Windows Approach by a more or less playful application within a framework of simple assumptions and models. This simple scenario has been formulated by the German Advisory Council on Global Change for its special report for the Conference of the Parties in Berlin 1995 (WBGU, 1995a and 1995b).

The WBGU scenario starts from two basic principles which guide the specification of the tolerable window in the impact space. The Council formulated these principles as follows:

" 1. the preservation of Creation and

2. the avoidance of unacceptable costs." (WBGU, 1995:7)

These two principles represent the criteria for the tolerance of climate-induced stresses for man and environment. The expression of these criteria in terms of geophysical and societal parameters and the translation of these parameters into climatic parameters yields the tolerable window of climate evolution (plain 3 in Fig. 1). Within the simple scenario used here, the first principle is specified by the consideration of the global mean temperature in recent history, i.e. the recent Quaternary. We require that the future temperature does not exceed the extreme temperatures within this period too much. Taking the figures of Schönwiese (1987) and adding a margin of 0.5 °C at each end of the temperature regime, we get a *domain from 9.9 °C to 16.6 °C.*

The second principle represents a simple socioeconomic constraint. It seems to be reasonable to expect the socioeconomic costs to be a function of the rate of temperature change (adaptation costs, not equilibrium!). In the case of equilibrium there exist a variety of economic assessments of damage costs, which calculate impact costs of CO_2 doubling to be about 1–2% of the global GNP (e.g. Nordhaus, 1991; Cline, 1992; Fankhauser, 1993). Assuming a corresponding temperature rise of about 2 °C at the end of the next century, we obtain 0.2 °C/decade as the average rate of temperature change. Although these damages seem to be rather moderate on the global level, it neither considers extreme events nor special regional aspects. As for the developing countries, the impact is on the order of 3%; the effects for the formal sector can increase the damages well beyond 5%. Therefore we can consider this type of damage as being on the edge and we thus reformulate the second principle as *"the avoidance of rates of temperature change larger than 0.2 °C/decade".* Since the ability of adaptation is assumed to decrease at higher temperature, however, the acceptable rate of temperature change should tend towards zero when approaching the upper limit of the temperature domain.

Using these assumptions and a consistent – yet somewhat arbitrary – mathematical formulation of the different constraints, we obtain the effective tolerable window $\mathcal{D}$ of future climates which is shown in Figure 2. Together with the boundary $\mathcal{B}$, the shaded area represents the admissible climatic conditions. The strange-looking positive and negative singularities at T = 9.9 °C and

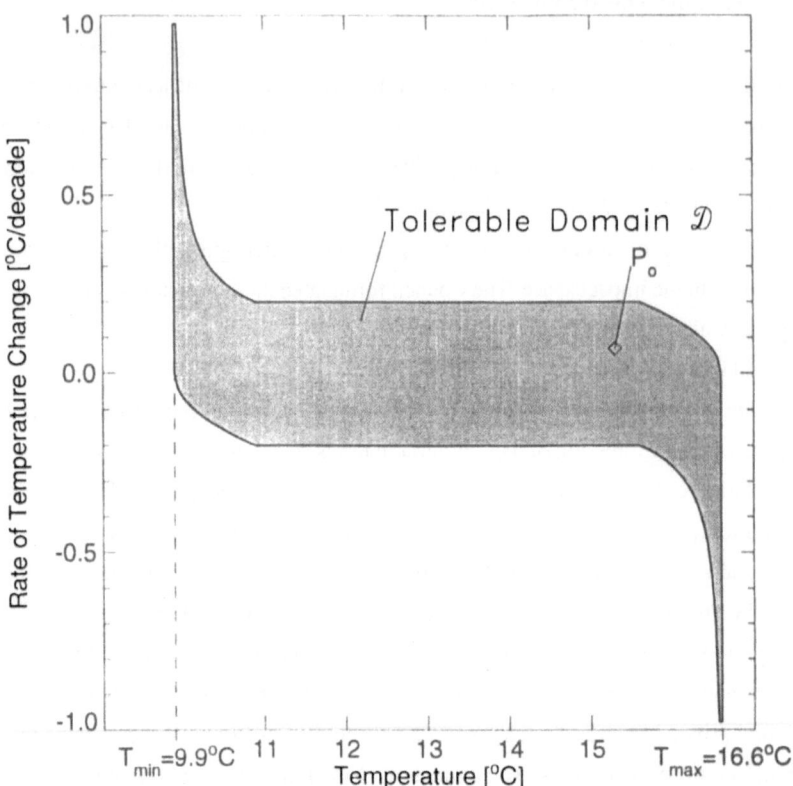

Figure 2. The domain of tolerable climate evolution expressed by the temperature and the rate of temperature change. This "climate window" is estimated (a) from the historical global mean temperatures in the recent Quaternary and (b) from simple arguments concerning potential climate impacts on the world economy. Any future climate evolution has to stay within this domain. The point P_0 denotes the status quo.

$T = 16.6$ °C respectively, result from the symmetric mathematical formulation of $\mathcal{D}$. Yet simple considerations of the dynamics in T and $\dot{T}$ show immediately that these "tails" can hardly be reached. In terms of Figure 1 this domain represents both the window in the impact-relevant geophysical space and the climate evolution space (note the similarity of the window shapes). The point P_0 ($T_0 = 15.3$ °C, $\dot{T} = 0.07$ °C/decade) marks the status quo as it results from temperature and concentration measurements which are fed into the climate model described below.

In order to transform the specified domain of tolerable climates into a domain of admissible GHG emissions, we have to make use of dynamic models for climate and GHG concentrations. We have used the following model from Hasselmann (1995, private communication) which is restricted to CO_2 for the sake of simplicity:

$$\dot{F} = \beta E$$

$$\dot{C} = BF + \beta E - \sigma C \qquad\qquad \text{[1. a – c]}$$

$$\dot{T} = \mu \ln \frac{C + C_1}{C_1} - \alpha(T - T_1)$$

Here, E denotes the global annual emission of CO_2 in Gt C/a, F is the cumulative anthropogenic emission in ppm, C is the CO_2 concentration in the atmosphere, and T denotes the global mean temperature. The concentration is measured relative to the pre-industrial levels, i.e. $C_1 = 290$ppm. The pre-industrial temperature is denoted by T_1. The parameters – except β which is a physical parameter (conversion from Gt to ppm$= 0.47$ ppm/Gt) – had been determined by comparison either with real world data or with results from GCMs. The values are shown in Table 1 together with the initial conditions for each variable.

Table 1. Values of the parameters and initial conditions in the combined carbon cycle climate model of Equations [1.a–c]

Parameter	Values and units	Variable	Initial value
B	0.0032 a^{-1}	E	7.9 Gt C/a
σ	0.022 a^{-1}	F	200 ppm
μ	0.087 °C/a	C	70 ppm
α	0.017 a^{-1}	T	15.3 °C
T_1	14.6 °C		
C_1	290 ppm		

In principle, we are now in a position to compute the admissible window of CO_2 emission profiles. Unfortunately, this simple model is already far too complicated to obtain an easy analytical characterization of this domain. We thus immediately jump to the next step, i.e. we evaluate the admissible emission paths by a superior criterion and directly derive the window shape in the policy space which already represents the result of the intersection procedure described in Section 2. In the following two sections, we use two different types of these criteria: optimization and risk avoidance. In these cases, we may use well-known mathematical principles and tools which do not require explicit knowledge of all admissible profiles.

4. Optimization of cumulative emissions

In general, the fundamental strategy of optimization requires the maximization of an appropriately-defined functional within a given time horizon. This functional, expressed in terms of the system variables, is chosen to represent an accepted measure of societal benefits. For example, economic cost-benefit assessments try to maximize the monetized or GNP-related macroeconomic benefits. As the total benefit in general will be a "function of the emission as a function of time", it is from the mathematical point of view a functional rather than a function.

Yet our goal is different, as we want to specify all those emissions which are compatible with the climate window. In order to characterize this set, we address the following question: *what is the maximal amount of CO_2 which can be released without jeopardizing the worlds' climate?* This corresponds to the maximization of the cumulative emissions F within a certain time horizon. This set depends on the types of possible emission profiles. As a first very crude property, we only specify a lower and an upper bound of emissions, $E_{min}=0$ and E_{max}, respectively.

We now can express the task of optimization in terms of mathematical control theory (Pontrjagin et al., 1964; Cesari, 1983):

> *Consider a specific time horizon τ. Search for a control function E(t), $t \leq \tau$ such that*
> *(1) the corresponding climate evolution, computed by Equations [1.a–c], stays within the climate window according to Figure 2 at all times,*
> *(2) the cumulative emission $F(\tau) = \int_0^\tau E(t)\, dt$ takes a maximum.*

Although this states the problem in its general form, there are two ingredients which make things more complicated. Firstly we had to introduce the time horizon τ and secondly we require the climate to stay within the domain $\mathcal{D}$ at *all* times, although we want to control it up to time τ only. The seeming contradiction within the latter statement can be resolved by the modified requirement that the control up to time τ should *in principle* allow the climate to stay within the domain for all later times. Therefore we modify the domain in such a way that it only contains those points for which a positive-semidefinite emission profile exists which keeps the climate evolution tolerable for all times (there are many mathematical subtleties of the existence of such emission profiles on which we do not want to elaborate). Henceforth, when talking about the climate window, we always refer to this modified domain. In practice, however, the relevant climate evolution will never reach the newly restricted regimes.

Concerning the first point, i.e. the introduction of a specific time horizon, it has to be stated that there is no preference for one or the other τ. We thus compute the maximal cumulative emission F_{max} as a function of τ. In principle the cumulative emissions for different time horizons might be achieved by different control functions, i.e. emission paths. Again practice turns out to be quite satisfying: there is *one* single emission profile which maximizes F for *all* time horizons. We now can reformulate the problem and turn it into a "velocity optimal" problem. Solution methods for this type of problems are well established and particularly simple (at least compared to the general optimization problem stated above).

4.1 Mathematics of optimal control

It can easily be seen that the function $F_{max}(\tau)$ increases monotonously. Thus one can chose an arbitrary F and look for that control function which minimizes the time τ to reach this specific amount of cumulative emissions. By varying F, we obtain a function $\tau_{min}(F)$ which is the inverse of the function searched for. In order to find these solutions one can make use of Pontrjagins Principle (Pontrjagin et al., 1964) specifying the necessary properties of an optimal solution. The actual proof by contradiction shows that any trajectory staying in the inner regime of $\mathcal{D}$ is suboptimal as it does not fulfill these conditions. The starting point for the control theoretical analysis is the formulation of a Hamiltonian H by introducing the auxiliary variables ψ_i, $i=1,2,3$. Then inside $\mathcal{D}$, the Hamiltonian takes the form

$$H(F,C,T;E) = \sum_{i=1}^{3} \psi_i f_i = \beta E(\psi_1 + \psi_2) + h(F,C,T,\psi_2,\psi_3), \qquad [2]$$

where the f_i for $i=1,2,3$ are given by the right hand sides of Equations [1.a–c], respectively. We have splitted the Hamiltonian into two parts, where the functional $h(F,C,T,\psi_2,\psi_3)$ represents the 'junk term' not required furthermore. Further we have

$$\frac{d\psi_i}{dt} = -\frac{\partial H}{\partial x_i}, \quad \bar{x} = (x_1,x_2,x_3) = (F,C,T). \qquad [3]$$

If the specific solution under control $\hat{E}$ is optimal, Pontrjagin's Principle requires H to take its supremum for an optimal control function $\hat{E}$. Thus we have

$$\hat{E} = \begin{cases} 0 & \text{if } \psi_1 + \psi_2 < 0 \\ E_{max} & \text{else} \end{cases} \qquad [4]$$

By skipping the details, one can give a more or less hand-waving argument for the type of velocity optimal solution. The argument runs as follows: consider a fixed cumulative emission F. Then it is clear that whenever the solution is velocity optimal and thus Equation [4] is fulfilled, one has to have $\hat{E}=E_{\max}$ as otherwise the emission F would have been reached at earlier times which is in contradiction to the assumption of velocity optimality. Therefore one has to start with $\hat{E}=E_{\max}$. If, however, this solution reaches the boundary, one has two options: stay at the boundary by emitting the corresponding levels of CO_2 or stopping all emissions, i.e. $\hat{E}=0$ which brings the trajectory back into the domain. In order to be part of the optimal strategy, the emission path followed by the latter option has to be switched to $\hat{E}=E_{\max}$ again where we have used the first argument given above. Therefore any optimal solution lying inside the domain has to have at least one switch in signs of $\psi_1 + \psi_2$.

Now, one has to make use of the so-called *transversality principle* which specifies the form of Pontrjagin's Principle for open final points (Pontrjagin et al., 1964), i.e. unspecified final concentration C and final temperature T in our case. The statement is basically that $\psi_2^1 = \psi_3^1 = 0$ where the superindex 1 denotes the value at the final point of the trajectory. Inserting this constraint into Equation [3] one can easily obtain $\psi_2(t) = 0, \psi_1(t) = const.$ which immediately implies that *no change of signs and therefore no switch from $\hat{E}=0$ to $\hat{E}=E_{\max}$ is possible*. Thus there exists no solution with more than the 'initial branch' within the domain which fulfills Pontrjagin's Principle. This implies that, whenever an optimal solution exists, it basically follows the boundary.

4.2 Realization of optimal solutions

Conclusively we can formulate the optimal strategy with respect to maximal cumulative emissions:

Get to the boundary of the tolerable window as fast as possible and then stay there.

According to the climate model Equations [1.a–c], the fastest way to reach the boundary is given by a delta-peak like emission of 124.3 Gt C. The successive control keeping the climate evolution on the boundary can easily be computed by simple numerical algorithms for the solution of ordinary differential equations. Figure 3 shows the corresponding emission profile (a), concentrations (b), and phase space dynamics (c) for the next 200 years. The cumulative emission within this time horizon is about 870 Gt C. However, because of the jumps within the emission profile which are due to the discontinuities in the boundary of the tolerable window, this strategy has to be considered as not feasible. Most dramatic are the δ-peak at the beginning and the 25% reduction within a single year. Therefore the realization of this profile would have unbearable economic and social side-effects, as no economy will endure a reduction of about 40% within a few years. Therefore the filter of nonacceptable quota and policies described in Section 2 forbids the choice of this particular profile.

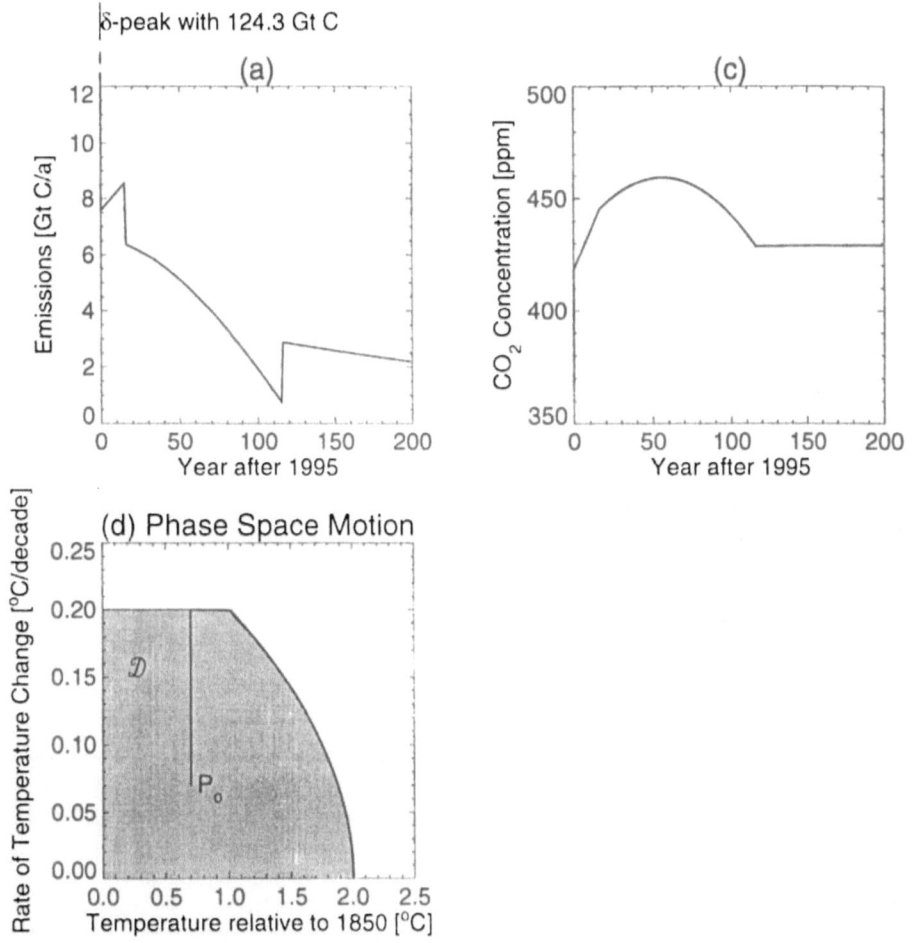

Figure 3. Evolution of the optimal solution with regard to the maximization of cumulative emissions within the next 200 years. As is typical for optimal control functions, distinct jumps within the emission profile – which actually represents the control function – can be seen (a). Besides an initial δ-peak emission of about 124.3 Gt, two further jumps at times $t = 15a$ and $t = 116a$ are detected. These jumps lead to discontinuities in the concentration profile which stabilizes at about 430 ppm (b). The solution is specified as the one which runs to the boundary of the domain $\mathcal{D}$ and stays there at all later times (c).

By the restriction of possible emission profiles to those which have high degrees of continuity and smoothness, an alternative emission profile is obtained which is shown in Figure 4. After a transition period of six years, the emissions are reduced at an annual rate of ca. 1% till the year 2155. Afterwards the reduction rate can be decreased to 0.25%. In this case the cumulative emission for $\tau = 200a$ is about 802 Gt C. This strategy can be considered as feasible as well as

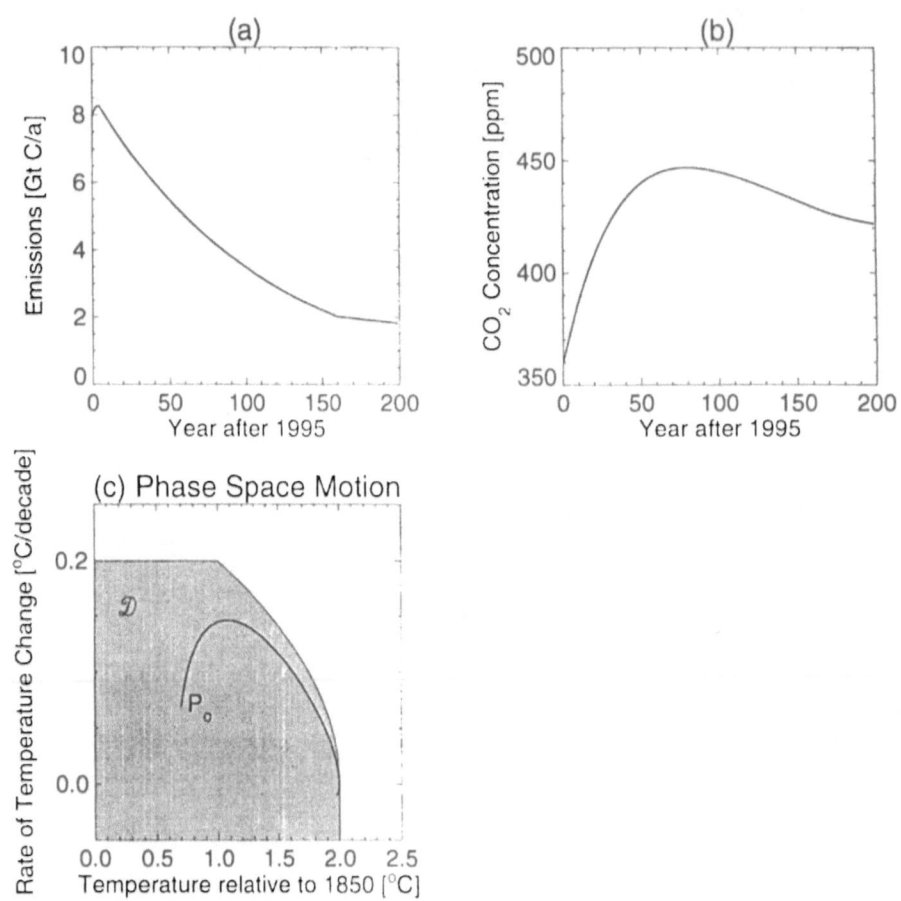

Figure 4. Same as Figure 3 but now for the optimal solution for smooth control functions. The time paths of the emissions, the concentration and the climate are smooth rather than characterized by a number of jumps.

acceptable from the benefit point of view (still very large cumulative emissions) and therefore lies well within the domain of acceptable policies. Furthermore it provides a high degree of planning certainty as the constant rate of reduction for about 150 years is a good baseline for the formulation of policy instruments. For examples of possible national allowances based on this strategy see WBGU (1995b).

So far we have not carried out the successive steps of the entire TWA as they were explained in Section 2. Nevertheless the exemplary analysis described should clarify most of the methodological elements of this new approach.

5. Including uncertainties: Risk analysis

Both strategies discussed in the previous section had been subject to the optimization criteria in terms of cumulative emissions and benefit. Due to the convex shape of the tolerable window and the monotonous dependence of the climate on the emissions, the corresponding climate evolution runs quite close to the boundary of the tolerable window. We have to take into account, however, that due to various uncertainties the strategies nevertheless might lead to intolerable climate states. Therefore a risk analysis has to be carried out.

In general the analysis of risks has to take into account different levels of damage together with the specific probabilities of their being achieved. Within the tolerable window approach things become much easier as we have a binary damage function: a specific climate evolution is considered to be acceptable or not. Thus, we can define the risk as the *probability of reaching nonacceptable climates* which has to been computed with respect to the chosen strategy $\tilde{E}$. In principle, three different aspects of uncertainty are involved:

(1) The uncertainty of planning, i.e. the intended or unintended failure of a chosen strategy. This leads to the time-dependent planning induced risk $R_{\tilde{E}}^{P}(t)$, i.e. the risk of having followed an unacceptable climate evolution within time t.

(2) The uncertainty of tolerance, i.e. the tolerable window might not correctly reflect the acceptance or nonacceptance of the climate evolution. The corresponding window-induced risk is denoted as $R_{\tilde{E}}^{W}(t)$.

(3) The uncertainty within the climate system, i.e. knowledge and modeling uncertainties. This induces the risk $R_{\tilde{E}}^{C}(t)$.

One can formulate exact mathematical expressions for each component as well as for the overall risk $R_{\tilde{E}}(t)$ (Petschel-Held, 1997). The latter two aspects differ from the first one by the type of uncertainty. Whereas for (2) and (3) the uncertainty is basically due to incomplete knowledge and therefore can be considered as exogenous, the uncertainty of planning is endogenous, i.e. it cannot be resolved but is immanent to the system.

Here we want to restrict ourselves to the analysis of the first two uncertainty aspects. In order to compare different emission strategies, we have investigated profiles which are classified by two parameters τ_w and r, referring to the time following the IPCC "business as usual" Scenario (IPCC, 1990) and the successive rate of annual emission reduction, respectively, i.e.

$$\tilde{E}(t) = \begin{cases} E_0(1+0.017 \cdot t) & \text{if } t \le \tau_w \\ E(\tau_w) \cdot \exp[-r(t-\tau_w)] & \text{else.} \end{cases} \qquad [5]$$

The window uncertainty was modeled by the variation of the thresholds for acceptable rates of temperature change ($\dot{T}_{max} \in [0.15, \ 0.25]$ °C / decade) and temperature ($T \in [16.6, \ 16.9]$ °C) with equal probabilities. The planning uncertainty was taken into account by a stochastic diffusion process of the rate of emission change overlaid on its strategy-specific time evolution Equation [5]. This leads to an ensemble of emission profiles with linearly increasing variance.

Figure 5a–c shows some isolines of the computed risks at $t = 150$ years, i.e. for the year 2145. Note that this also includes all orbits which have left the domain of tolerance at earlier times. The waiting time τ_w is plotted along the abscissa [years] and the reduction rate r along the ordinate [%]. Thus each point in the diagram denotes a single strategy. Figure 5(a) refers to the planning induced risk $R_{\dot{E}}^P(t)$, Figure 5(b) to the risk $R_{\dot{E}}^W(t)$ due to window uncertainties, and Figure 5(c) presents the results for the combined risk $R_{\dot{E}}^{WP}(t)$.

It can be seen that the general structure of risks looks the same for all types of risk: in order to have a low risk, a substantial reduction with rates of at least 0.5% has to be launched quite soon. The longer the implementation of reduction measures is delayed, the stronger the reduction will have to be. Moreover, this relation is superlinear, i.e. the necessary reduction rate grows faster than linear in the time of its implementation. Computations with other climate models reveal this general structure (Petschel-Held, 1997). Furthermore, one obtains the risks of the optimal strategy obtained in Section 4.2 which roughly is given by $\tau_w = 6$ years and $r = 1.0\%$ as $R_{\dot{E}}^P = 62\%$, $R_{\dot{E}}^W = 13\%$, $R_{\dot{E}}^{WP} = 41\%$. Altogether, these results suggest to start with an effective climate protection strategy as soon as possible.

6. Conclusions

Any profound integrated analysis of climate protection strategies has to start from a careful specification of the interfaces between normative settings (political issues) and system theoretical considerations (scientific issues). In this paper, we have presented an approach which is based on the specification of these interfaces in terms of tolerable domains for *climate impacts*, *policy options*, and *instruments* for the realization of reduction measures. By the analysis of the interdependencies between the relevant subsystems of the entire Earth System, it is possible to derive the set of political options which are tolerable with regard to *all* three aspects of normative settings. The necessary system analysis makes use of an inverse mapping strategy between the different subsystems, e.g. by mapping the domain of tolerable climate impacts into the climate evolution subspace.

We have illustrated this approach by a simple example based on a scenario developed by the German Advisory Council on Global Change (WBGU). The example starts from the specification of a simple domain of tolerable climate evolution in terms of temperature and rate of

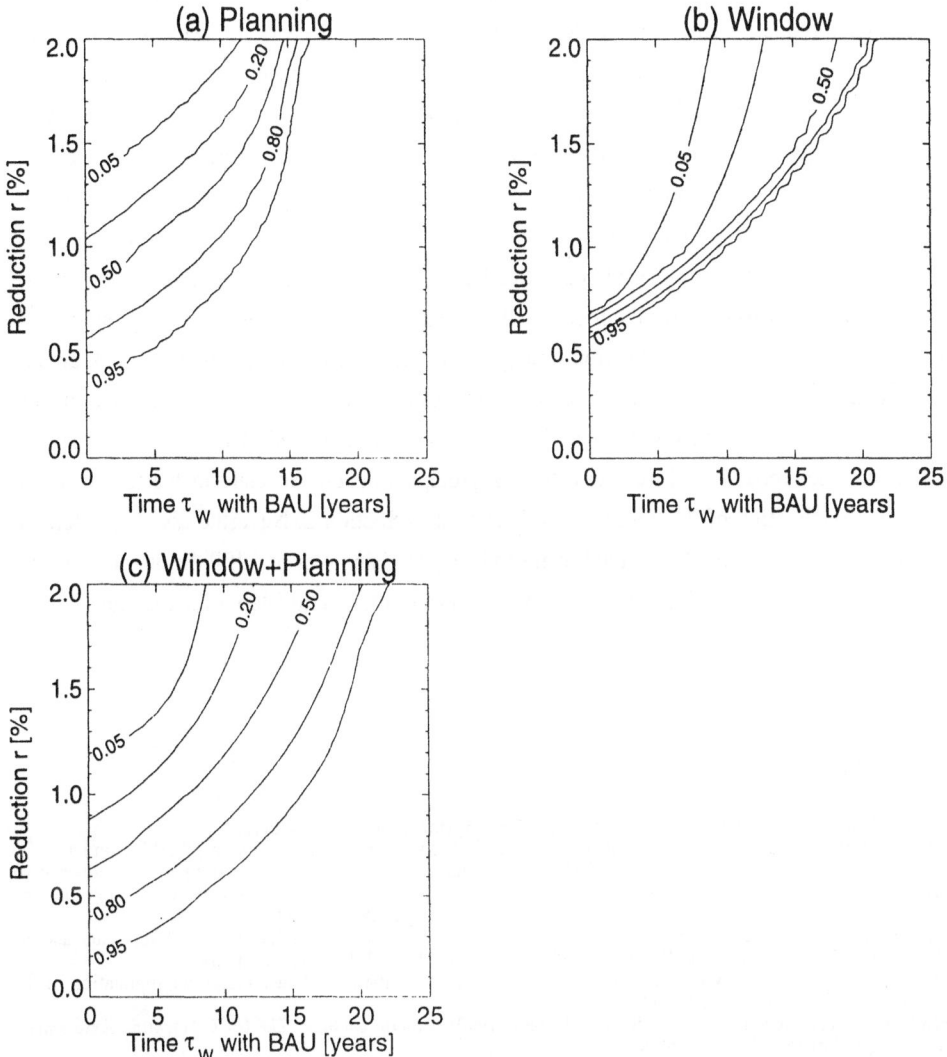

Figure 5. Isolines for the computed risks of following an unacceptable climate evolution. Each point in the plains represents a distinct emission reduction strategy (see Eq. [5]).
(a) represents the risk due to planning uncertainty, i.e. to some type of failure of the chosen strategy. In extension the uncertainty in the choice of the climate window is included and the risk thus induced is shown in (b). Finally both uncertainties are taken into account to obtain the combined risk presented in (c). The undulation of the curves is due to the discrete scan of the parameter space (τ_W, r).

temperature change. Two superior criteria have been used to evaluate those future emission pro-files which keep the climate evolution within this domain of tolerance: maximization of cumulative emissions and limitation of uncertainty induced risks. Both investigations suggest that an effective emission reduction should be started within the next 5 to 10 years at most, e.g. by reducing global emissions at an annual rate of about 1%. Such measures make it possible to have large benefits in terms of high cumulative emissions as well as high security, both with regard to planning certainty and with respect to low risks for leaving the domain. It turns out that the degree of difficulty in satisfying both criteria grows in a superlinear way according to the delay in implementation of reduction measures.

Although the simple example studied in Sections 3–5 shows some interesting results in itself, it will be necessary to improve the theoretical and empirical basis. In order to carry out a well-based assessment of climate protection strategies one has to take into account to a much greater extent details of the socio-economic subsystem, e.g. refined levels of tolerance, integrated models of the global economy, or region-specific climate sensitivities, and of the climate-biosphere system, e.g. the probability of weather extremes, precipitation, etc. Special care has to been taken on the regionalization of the approach as discussed in Section 2. Nevertheless, the Tolerable Windows Approach (TWA) presented seems to be a promising way to fulfill the duty of science: to give good and profound advice to policy makers in order to develop effective and positive climate protection measures.

References

Cesari, L (1983) *Optimization – Theory and Applications*. Springer-Verlag, New York.

Cline, W (1992) *The Economics of Global Warming*. Institute for International Economics, Washington DC.

Edmonds, J, Barns, DW and Ton, M (1993) The regional costs and benefits of participating in alternative hypothetical fossil fuel emissions reduction protocols. *In*: Y Kaya, N Nakičenovič, WD Nordhaus, and FL Toth (eds): *Costs, Impacts, and Benefits of CO_2 Mitigation*. CP-93-2, IIASA, Laxenburg.

Fankhauser, S (1993) The economic costs of global warming. *In*: Y Kaya, N Nakičenovič, WD Nordhaus, and FL Toth (eds): *Costs, Impacts, and Benefits of CO_2 Mitigation*. CP-93-2, IIASA, Laxenburg.

Hope, C, Anderson, J and Wenman, P (1993) Policy analysis of the greenhouse effect: an application of the PAGE model. *Energy Policy*, 21:327.

IPCC (Intergovernmental Panel on Climate Change) (1990) *Climate Change: The IPCC Scientific Assessment*. Cambridge University Press, Cambridge.

IPCC (Intergovernmental Panel on Climate Change) (1996) *Climate Change 1995: Economic and Social Dimensions of Climate Change*. Cambridge University Press, Cambridge.

Kaya Y, Nakičenovič, N, Nordhaus, WD and Toth, FL (eds) (1993) *Costs, Impacts, and Benefits of CO_2 Mitigation*. CP-93-2, IIASA, Laxenburg.

Nakičenovič, N, Nordhaus, WD, Richels, R and Toth, FL (eds) (1994) *Integrative Assessment of Mitigation, Impacts, and Adaptation to Climate Change*. CP-94-9, IIASA, Laxenburg.

Nordhaus, W (1991) To slow or not to slow: the economics of the greenhouse effect, *Econ. J.* 101:920.

Petschel-Held, G (1997) Long-term perspectives of short-term climate protection strategies. Manuscript. PIK, Potsdam.

Petschel-Held, G, Schellnhuber H-J, Bruckner T, and Hasselmann K (1997) The Tolerable Windows Approach: an inverse integrated assessment of climate change. Mimea, PIK: Potsdam.

Pontrjagin, LS, Boltjanskij, VG, Gamkredlidze, RV and Miscenko, EF (1964) *Mathematische Theorie optimaler Prozesse*. Oldenbourg Verlag, München.

Richels, R and Edmonds J (1994) The economics of stabilizing atmospheric CO_2 concentrations. *In*: N Nakičenovič, WD Nordhaus, R Richels and FL Toth (eds): *Integrative Assessment of Mitigation, Impacts, and Adaptation to Climate Change*. CP-94-9, IIASA, Laxenburg.

Schönwiese, C-D (1987) Climate variations. *In*: D Etling, M Hantel, H Kraus and C-D Schönwiese (eds): *Landolt-Boernstein – Zahlenwerte und Funktionen aus Naturwissenschaft und Technik*. Neue Serie, Gruppe V, 4th volume, Springer-Verlag, Heidelberg.

Toth, FL (1994) Practice and progress in integrated assessments of climate change: a review. *In*: N Nakićenović, WD Nordhaus, R Richels and FL Toth (eds): *Integrative Assessment of Mitigation, Impacts, and Adaptation to Climate Change*. CP-94-9, IIASA, Laxenburg.

Toth, FL, Bruckner, T, Füssel, H-M, Leimbach, M, Petschel-Held, G and Schellnhuber, H-J (1997) The tolerable windows approach to integrated assessments. Proceedings of the IPCC Asia-Pacific Workshop on Integrated Assessment Models, Tokyo, 10–12th March 1997, (forthcoming).

UN (United Nations) (1992) *United Nations Framework Convention on Climate Change*. Convention Text. IUCC, Geneva.

UN (1995) *Report of the Subsidiary Body for Scientific and Technological Advice on the work of its first Session*. FCCC/SBSTA/1995/3.

WBGU (German Advisory Council on Global Change) (1995a) *Scenario for the derivation of global CO_2 reduction targets and implementation strategies*. Statement on the occasion of the First Conference of the Parties to the FCCC in Berlin. WBGU, Bremerhaven.

WBGU (1995b) *Welt im Wandel: Wege zur Lösung globaler Umweltprobleme. Jahresgutachten 1995*. Springer-Verlag, Berlin.

Wigley, TML, Richels, R and Edmonds, JA (1996) Economic and environmental choices in the stabilization of atmospheric CO_2 emissions. *Nature* 379: 240.

Appendix 1
List of participants

Joe Alcamo
Wissenschaftliches Zentrum für Umweltsys-
temforschung, Gesamthochschule Kassel
Kurt-Wolters-Strasse 3
D-34109 Kassel, Germany

Gerhard Berz
Münchener Rückversicherung, Abt. Sa/XL-EL
Königinstrasse 107
D-80802 München, Germany

Jürgen Blank
Westf. Wilhelms Universität Münster
Lehrstuhl für Volkswirtschaftstheorie
Universitätstrasse 14–16
D-48143 Münster, Germany

Wolfgang Cramer
Potsdam-Institut für Klimafolgenforschung
e.V.
P.O. Box 601203
D-14412 Potsdam, Germany

Jae Edmonds
Batelle Pacific Northwest Laboratories
901 D Street, S.W., Suite 900
Washington, DC 20024-2115, USA

Michael Ernst
Bundesministerium für Umwelt
Bernkasteler Str. 8
D-53175 Bonn, Germany

Klaus Hasselmann
Max-Planck-Inst. f. Meteorologie
Bundesstr. 55
D-20146 Hamburg, Germany

Peter Hennicke
Wuppertal Institut
Döppersberg 19
D-42103 Wuppertal, Germany

Harald Jacobson
Center for Political Studies, Room No. 465
The Institute for Social Research
University of Michigan
Ann Arbor, MI 48106-1248, USA

Eberhard Jochem
Fraunhoferinstitut, ISI
Breslauer Str. 48
D-76139 Karlsruhe, Germany

Julia M. Kundermann
BMBF - Forschungsministerium, Ref 422
Heisemannstr. 2
D-53170 Bonn, Germany

Hans-Jochen Luhmann
Wuppertal Institut
Döppersberg 19
D-42103 Wuppertal, Germany

Nebojsa Nakičenovič
International Institute for Applied Systems
Analysis
Schlossplatz 1
A-2361 Laxenburg, Austria

Gerhard Petschel-Held
Potsdam-Institut für Klimafolgenforschung
e.V.
P.O. Box 601203
D-14412 Potsdam, Germany

Klaus Rennings
Zentrum für Europäische
Wirtschaftsforschung
P.O. Box103443
D-68034 Mannheim, Germany

Richard Richels
Energy Analysis and Planning Dept.
Electric Power Research Institute
3412 Hillview Ave.
Palo Alto, CA 94303, USA

Rolf Sartorius
Umweltbundesamt (UBA)
Fachgebiet Schutz der Erdatmosphäre
Bismarckplatz 1
D-14193 Berlin, Germany

Hans-Joachim Schellnhuber
Potsdam-Institut für Klimafolgenforschung
e.V.
P.O. Box 601203
D-14412 Potsdam, Germany

Manfred Stock
Potsdam-Institut für Klimafolgenforschung
e.V.
P.O. Box 601203
D-14412 Potsdam, Germany

Ferenc L. Toth
Potsdam-Institut für Klimafolgenforschung
e.V.
P.O. Box 601203
D-14412 Potsdam, Germany

David Victor
International Institute for Applied Systems
Analysis
Schlossplatz 1
A-2361 Laxenburg, Austria

Gary W. Yohe
Department of Economics
Wesleyan University
Middletown, CT 06457, USA

Wolfgang Zuckschwerdt
Projektträger Umweltsystemforschung
(PT/USF)
Godesberger Allee 117
D-53175 Bonn, Germany

Appendix 2
Workshop Program

Cost-Benefit Analyses of Climate Change
Joint PIK/WI Workshop, November 10–11, 1995, Potsdam, Germany
Financially supported by BMBF

DAY 1:   November 10

08:45   Opening – *Hans-Joachim Schellnhuber*
09:00   Introduction, overview, objectives – *Ferenc Toth*

Session 1: Regional climate sensitivity, impacts, and adaptation
          (Chair: *Hans-Joachim Schellnhuber*)

09:20   Theme 1: Criteria for evaluating climate impacts: Regional differences in impacts
                    and adaptation options
                    Jae Edmonds

10:00   Theme 2: Climate sensitivity 1: Ecosystem damages
                    Wolfgang Cramer, Joe Alcamo

10:50   Break

11:10   Theme 3: Climate sensitivity 2: Monetary and non-monetary social damages
                    Gerhard Berz, Klaus Rennings

12:00   Theme 4: Deriving GHG reduction targets from Article 2
                    Klaus Hasselmann, Hans-Jochen Luhmann

Session 2: Emissions reduction instruments and costs
          (Chair: *Peter Hennicke*)

14:00   Theme 1: International burden sharing: criteria and their implications on regional
                    emission paths
                    Nebojsa Nakičenovič, Jae Edmonds

14:50   Theme 2: European and OECD stabilization targets: What they bring, how much
                    do they cost
                    Richard Richels, Eberhard Jochem
15:40   Break

16:00   Theme 3: Economic efficiency and optimality as policy criteria
                    Harald Jacobson

16:50   General discussions and summary of Sessions 1 and 2

DAY 2:   **November 11**

Session 3: **Political context and policy implementation**
            (Chair: *Julia Kundermann*)

09:00   Theme 1: Controlling prices vs. quantities nationally and internationally: tradable
                     permits, taxes, JI
                     Gary Yohe (paper summarized by F. Toth), Jürgen Blank

09:50   Theme 2: Possible/expected efficiency of any climate policy
                     David Victor, Michael Ernst

10:40   Break

11:00   Theme 3: Reverse integration: the tolerable window approach
                     Gerhard Petschel-Held

11:50   Session 3 Summary

12:10   **Final discussions and workshop summary**
            Chair: *Hans-Joachim Schellnhuber*

14:00   Adjourn

Subject Index

Swiss Priority Programme
Environment Synthesis

Life Cycle Assessment (LCA) – Quo v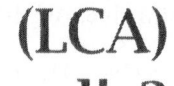?

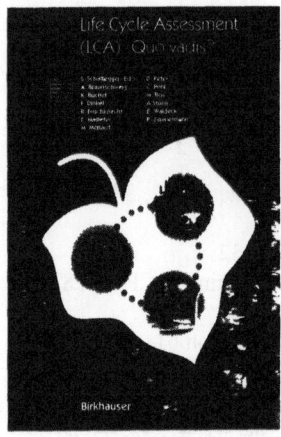

Edited by **S. Schaltegger**, *Center of Economics and Business Administration, University of Basel, Switzerland*
1996. 168 pages. Softcover
ISBN 3-7643-5341-4

LCA - Quo vadis? discusses overarching topics, new developments and major problems of Life Cycle Assessment (LCA), and compares LCA with site-specific environmental management. The text profits from two years of interdisciplinary, coordinated research activities of the Priority Programme Environment of the *Swiss National Science Foundation*. Furthermore a case study applying LCA to the product life cycle of beer (Feldschlösschen Ltd.) is presented. A major part of the book deals with the following LCA related topics:

√ **How should system boundaries of a product life cycle be drawn?**

√ **How can environmental interventions be allocated to products?**

√ **How are background inventory data collected and used?**

√ **How can imprecision in the LCA method be ascertained and checked?**

√ **How can relevant environmental interventions be distinguished from irrelevant ones?**

√ **What requirements should a software tool for LCA meet?**

A concept of site-specific LCA is proposed in response to criticism of the current approach of LCA. Furthermore, managerial eco-controlling - the emerging method of site-specific environmental management - is discussed. The book concludes with an outlook of possible paths in the future development of LCA.

Birkhäuser Verlag • Basel • Boston • Berlin

Swiss Priority Programme
Environment Themenhefte

**Environmental Policy
Between Regulation and Market**

Enviro⸻tal
Policy between
Regulation
and Market

Edited by **C. Jeanrenaud**,
University of Neuchâtel, Switzerland
1997. 376 pages. Softcover
ISBN 3-7643-5319-8

Environmental policies have traditionally relied on direct controls and on government investment to protect natural resources. Today, the drawbacks and impediments to this approach are evident: heavy burdens borne by companies and the community, complex regulations, a danger of legislative inflation, difficulties in meeting the goals set, to name a few. In response, the environmental authorities in many countries have begun to reassess the efficacy of their programs, with the result that market incentives and voluntary agreements with companies or branches of industry have been added to the arsenal of traditional environmental protection measures.

There are great expectations for new economic instruments, which offer the twofold advantage of giving companies more freedom in the choice of means, and of increasing the chances for meeting goals in a more cost-effective way. The authors of this book analyse these instruments - green taxes, tradeable permits, covenants, joint implementation, internationally tradeable quotas - from the point of view of cost-effectiveness, their ability to achieve environmental goals, and public and corporate acceptability. They endeavour to determine on the basis of experience to date, whether these instruments are living up to the hopes placed in them.

Birkhäuser Verlag • Basel • Boston • Berlin

Schwerpunktprogramm Umwelt Synthesebuch

Mut zum ökologischen Umbau

Innovationsstrategien für Unternehmen, Politik und Akteurnetze

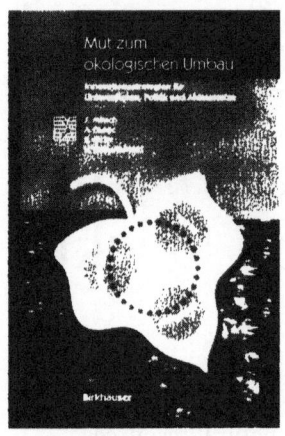

von J. Minsch, A. Eberle, U. Schneidewind, *Institut für Wirtschaft und Okologie, Universität St. Gallen / B. Meier, Geographisches Institut der Universität Bern*
1996. 296 Seiten. Gebunden.
ISBN 3-7643-5320-1

Eine nachhaltige Wirtschaftsweise steht in den Industrieländern noch aus - trotz zahlreicher Umweltinnovationen in Politik und Wirtschaft. Die Autoren dieses Synthesebuches aus dem SPP Umwelt des Schweizerischen Nationalfonds gehen den Ursachen nach, zeigen wo die ökologische Innovationsoffensive in Unternehmen, in regionalen Akteurnetzen und in der Umweltpolitik heute steht und wie sie sich weiterentwickeln muß. Ihre Ergebnisse stützen sie auf ein interdisziplinäres Forschungsprojekt, in dem Volkswirte, Betriebswirte und Wirtschaftsgeographen am Beispiel der Schweiz Umweltinnovationen untersucht haben. Das Buch verbindet reichhaltige empirische Daten mit einer umfassenden konzeptionellen Analyse und zeigt: ökologischer Wandel kommt weder alleine von unten noch von oben. Er braucht vielmehr das geschickte Zusammenspiel von unternehmerischer Initiative, umweltpolitischer Rahmensetzung und geeignetem Netzwerkmangement.

Mut zum Umbau ist erforderlich.

Die wirtschaftliche Entwicklung geht in Richtung Nachhaltige Entwicklung. Umweltpolitische Lähmung prägt das Bild, die ökologische Situation verschlechtert sich und die Umsetzung der Beschlüsse von Rio werden verzögert.

Birkhäuser Verlag • Basel • Boston • Berlin